AF838217

Transportable Environments 3

Edited by Robert Kronenburg

Co-edited by Filiz Klassen

Taylor & Francis
Taylor & Francis Group
LONDON AND NEW YORK

First published 2006
by Taylor & Francis
2 Park Square, Milton Park, Abingdon, Oxon OX14 4RN

Simultaneously published in the USA and Canada
by Taylor & Francis Inc
270 Madison Ave, New York, NY 10016

Taylor & Francis is an imprint of the Taylor & Francis Group

© 2006 Robert Kronenburg and Filiz Klassen, selection and editorial material; individual chapters, the contributors

Typeset in Helvetica Light by Wearset Ltd, Boldon, Tyne and Wear
Printed and bound in Great Britain by the Alden Group, Oxford

All rights reserved. No part of this book may be reprinted or reproduced or utilized in any form or by any electronic, mechanical, or other means, now known or hereafter invented, including photocopying and recording, or in any information storage or retrieval system, without permission in writing from the publishers.

Every effort has been made to ensure that the advice and information in this book is true and accurate at the time of going to press. However, neither the publisher nor the authors can accept any legal responsibility or liability for any errors or omissions that may be made. In the case of drug administration, any medical procedure or the use of technical equipment mentioned within this book, you are strongly advised to consult the manufacturer's guidelines.

British Library Cataloguing in Publication Data
A catalogue record for this book is available from the British Library

Library of Congress Cataloging in Publication Data
Transportable environments III / edited by Robert Kronenburg, co-edited
by Filiz Klassen.— 1st ed.
p. cm.
Includes bibliographical references.
ISBN 0–415–34377–1 (pbk. : alk. paper) 1. Buildings, Portable. I.
Title: Transportable environments 3. II. Title: Transportable
environments three. III. Kronenburg, Robert. IV. Klassen, Filiz, 1965–
NA8480.T734 2005
721—dc22

2005010636

ISBN 0–415–34377–1

Transportable Environments 3

Contents

Technology

Projects

Illustration credits

Every effort has been made to contact the copyright holders of images found on the Internet without proper accreditation. The editors would like to thank Derek MacKay and Anna Sluszkiewicz for allowing them to borrow images from their extensive archives.

Alberto Pérez-Gómez: 1.1–1.4; Library of Congress American Memory Website: 2.1; SODRA Montréal © The Estate of Marcel Duchamp: 2.2, 2.3, 2.5, 2.6, 2.9; Austrian Frederick and Lillian Kiesler Private Foundation: 2.4, 2.7, 2.8, 2.10; Paul Cashman/ Dimitri Proano (Apple Is Design Group): 3.1; © Justin Beal: 4.1; © Derek MacKay: 4.4, 4.9; © Anna Sluskiewicz and 'ATS Notes' (atsnotes.com): 4.7, 4.8; © Xander Rombach: 4.3; Hades Publications Inc.: 6.2, 6.6; Dawes Collection: 6.3; © Chris Faust: 7.1–7.7; Hoberman Designs Inc.: 8.1–8.13; Sarah Bonnemaison: 9.1, 9.3–9.5; Christine Macy: 9.2; Keith Tufts: 9.6; © Marcin Padlewski 2004: 10.1–10.7; Andrew Furman: 11.1, 11.5; Peter MacCallum: 11.2; Mark Robbins: 11.3; Michael Heinrich: 11.4, 11.6–11.8; Meindert E. Versteeg: 12.1–12.7; Levitt Goodman Architects 2004: 13.1–13.6; Robert Kronenburg: 14.1; FTL Design and Engineering Studio: 14.4; Festo KG: 14.5; Lorenzo Appicella: 14.3; Branson Coates Architecture: 14.6; Future Systems: 14.7; Edmund Sumner/View: 15.1; Barry Halkin Photography: 15.2 (left); Elliot Kaufman Photography: 15.2 (right); Edward J. Hogan: 15.3; Panelite: 15.4–15.5; Marcel Wanders: 15.6; Alusion: 15.7; FTL Design Engineering Studio (photography by Michael Lent): 15.8–15.10; ONL (Oosterhuis Lénárd): 15.11; Society of Manufacturing Engineers, Fundamentals of Composites Manufacturing: Materials, Methods and Applications © 1989: 16.3; Jordan Brandt and Alejandro Ogata: 16.1–16.2, 16.4–16.9; Carolina Rodriguez and John Chilton: 17.1–17.11; © Neil Burford and Christoph Gengnagel: 18.1–18.11; Jenson Scientifics, LLC: 19.1b; University of Central Arkansas: 19.2; G. GIldea: 19.3a, 19.3b; Dr Jonathan Hare, The Creative Science Centre, Sussex University: 19.3c; F. Escrig: 19.4; H. Kobayashi, B. Kresling and J. F. V. Vincent, *Proceedings of the Royal Society*, London, B265, No. 1391: 19.5, 19.6; D. S. A. De Focatiis and S. D. Guest, Phil. Trans. the Royal Society, London, A360: 19.7; LOT-EK: 20.1–20.7; Fotoworks/Benny Chan, 1998: 21.1; Daniel Hennessy: 21.3; Robert Kronenburg: 21.5; Office of Mobile Design: 21.2, 21.4, 21.6, 21.7; Renzo Piano Building Workshop: 19.8; Santiago Calatrava: 19.9, 19.10; Matias Creimer: 23.1; Jessica Davies, Yvonne Cheng, Johanna Doukler and Tamsin Ford: 24.1; Rocco Sabalones Cetera 2004: 25.1; Piet Mazereeuw: 28.1; Janet McGaw: 29.1; Kaija Vogel (drawings) and Camellia Koo (photography): 30.1.

Foreword

This book results from the third conference in the Transportable Environments series, the objective of which is to examine the international range of research and practical activity in the field of portable architecture and its context. The first conference was held at the Royal Institute of British Architects in London in 1998, the second at the National University of Singapore in 2001. These locations were chosen because of their dramatically different geographic and cultural contexts – the concept of mobile building is undoubtedly a global one, but regional conditions can and do have dramatic effects on the form that architecture takes. By staging the most recent 2004 event in Toronto, Canada, the ambition was to attract delegates from diverse backgrounds to a new locality and to expand the opportunity to share current experiences affecting both local issues and lessons of global value.

This conference set out to explore the character and manifestation of the transportable environment, whether a building, a landscape or an interior space brought into existence in a specific location for a limited or specified period of time. The ambition was to present a public and academic forum where examples of research, design projects, buildings and cultural contexts requiring programmatic, constructional and design responses to evolving technological, social, economic and cultural conditions could be discussed. Because of the way in which the world is changing, flexible, transformable and transportable design is as important now as it was when, in past millennia, nomadic ways of life were dominant across the planet. Conference meetings are points in time, gathering diverse viewpoints and research opinions that stem from a specific historical moment in architectural development. A book establishes and communicates this taste of the variety of thought and practice and provides a more accessible resource for those who were not there. In that way it is rather like the architectural manifestation we set out to explore – ephemeral, contextually based – but, with this volume, having some lasting, useful and transportable impact.

The three keynote speakers in Toronto embody three important aspects of the field – theory, design and practical research. Dr Alberto Pérez-Gómez's work challenges the highest level of theoretical and historical architectural research. Ada Tolla and Giuseppe Lignano of LOT-EK are artists, architects and designers, whose work is at the cutting edge of innovation. Their projects and proposals challenge typical architectural design solutions by searching for new approaches. Chuck Hoberman is a creator of long-standing reputation responsible for tried and tested deployable structures that have found their way into all aspects of design culture from stadium-sized building to children's toys.

Complementing these speakers were a range of thematic papers, poster projects and performances. The aim of the conference was to gather research, knowledge and the experience of national and international experts with first-hand information on theoretical, educational, technological and design issues in a global and cross-disciplinary context. This book has its beginnings in that event but it is not simply a set of proceedings because each of the contributions that are included have been reviewed and updated so that they can be appreciated outside the context of the conference setting. In some cases project presenters have been asked to extend their work into a full essay, in others the emphasis has shifted or expanded to fall more directly into the themes of this publication. Like the previous *Transportable Environments* volumes, this work brings together a diverse range of viewpoints on many of the different manifestations this form of architecture takes. The study of transportable environments is a search for new interpretations of existing and innovative building technologies to fulfil requirements that cannot be met by the use of existing built forms. Though the essays within this work explain many interesting and important issues, they inevitably raise many more questions, making clear the enormous potential for further academic scholarship, research and design excellence that remains.

The editors must extend their gratitude to the keynote speakers, essay and project authors, and all the conference delegates who made Toronto such a memorable event; to Sarah Bonnemaison and Vladimir Krstic for their valuable counsel as members of the review committee; and to Caroline Mallinder and publishers Taylor & Francis for their continued support of this latest volume.

Robert Kronenburg, University of Liverpool, United Kingdom
Filiz Klassen, Ryerson University, Toronto, Canada

Theory, History and Context

Polyphilo's Thresholds: Alternatives for Nomadic Dwelling

Alberto Pérez-Gómez
McGill University

Over the centuries, nomadic life has offered humanity real possibilities for fulfilment, revealing to the individual a sense of participation in the cultural order and the natural world. Whether for the bedouin in the desert or the Christian pilgrim in the Middle Ages, enacting our condition of passerby upon the face of the earth by walking and belonging to no permanent physical place held a profound significance. The man-made structures that made this possible, such as the tent, the icon or the tabernacle, were imbued with genuine symbolism, touching all orders of life from the social and political to the religious. Our own project of mobility and contemporary cosmopolitanism only appears as new and revolutionary when seen against a civilization obsessed with permanence and predictability, a civilization that believes its destiny is to dominate and possess an external reality transformed into natural resources, while consciously or subconsciously concealing mortality and the ultimate ephemerality of all things human, from personal belongings to the powers of the mightiest empire. This dominant world-view is a relatively recent phenomenon.

It is important to recall that prior to the nineteenth century in Europe, much care was given to ephemeral structures in cities, such as triumphal arches, *tableaux vivants* and the like, built for specific religious or political celebrations. These singular events, lasting one or a few days, which revealed a political order through the transformed city, were universally significant for the inhabitants, conveying a sense of purpose and a general existential orientation. In a similar way to permanent architecture, ephemeral structures are capable of revealing a 'poetic image', one that affects us primarily through our vision and yet is fully sensuous, synaesthetic, thus capable of seducing and elevating us to understand our embodied consciousness's participation in wholeness. In addition, traditional architectural theories, which after Vitruvius privileged formal, monumental representations of geometrical order, always did so cognizant of the ephemeral condition of human life on earth, seeking to reconcile our human capacity for stability and precision that seemed to reflect the character of the eternal cosmos, with an openness to the gift of death.

During the last two centuries, our Western cultures have become increasingly obsessed with positive reason, excluding poetry and mythical stories as illegitimate forms of knowledge. Early democratic systems freed the individual from traditional, often oppressive, political orders, yet they also forced humanity to confront the abyss of nihilism. The reaction was a new and radicalized attachment to private property, to ethnic groups or nationalistic institutions, and to the production of historicist or technological architecture: ossified monuments with an inherent capacity to be transformed into ideological 'symbols', becoming potentially destructive and contributing eventually to the horrific genocidal wars of the twentieth century. It must be acknowledged that the political and epistemological revolutions of the eighteenth and nineteenth centuries contributed much to the physical freedom and comfort of the individual, the new voyeur and consumer of the metropolis. Yet they also tended to conceal the very nature of the human condition, its fundamental uncertainty, which, as a mediaeval Japanese poet once said, is what makes life wonderful and worth living.

Once Western humanity became fully modern, embracing the values of a progressive history between the eighteenth and the nineteenth centuries, the possibility of placelessness became real and, with it, nomadic life could become a potential nightmare. The genius of place, associated to the linguistic reality of traditional cultures, slowly vanished from our experience, progressively substituted by the isotropic, homogeneous space-time of technology. Seeking to solidify a historical reality into a stable and predictable ground for action, a crucial yet most difficult endeavour without the aid of the imagination, architects were often called to produce nostalgic monuments on the basis of ideological programmes such as liberalism, fascism, communism, all issuing from the ideals of the French Revolution. The result: historicist traces like the neo-Greek or Roman forms of Nazism, or the pseudo-scientific forms of international corporate architecture. Architecture embodies political values which tend to become repressive, and this has traditionally been implemented through heaviness and monumentality. Today, after the shortcomings of globalization have become more evident and our world struggles with its political and economic crises, it is only fair to seek alternatives.

It is crucial, nevertheless, that the project for a new nomadism be understood in the full context of our postmodern existence. In a true Nietzschean spirit we may believe that the strong values of traditional cultures, such as nationalism or ethnicity, should be weakened. Our architecture, however, must remain the repository of memories that allow for further action. This is crucial in a world increasingly reduced to a global village, in which religions and myths are no longer an intersubjective ground of being. It is not enough to resolve a technical problem of portability or sustainability, or simply to declare that the alternative to globalization is cultural nomadism in a post-traditional environment. We may abhor the monumental heaviness of old architecture, affirming as it did theocentric or false political values. Yet it is imperative that once we embrace 'lightness' as an alternative, we become fully aware of its implications, stepping outside a dialectic. The issue is not merely to oppose sedentary and historical to nomadic and ahistorical. As architects or designers we are called to envision a better future, our projects necessarily retain a utopian vector as part of their ethical dimension. Whether we design an ephemeral or a permanent structure, this ethical imperative is primary. In order to embrace it, we must understand that the ephemeral object must simultaneously offer a dwelling place and therefore, paradoxically, be *memorable*. And this is indeed the difficulty. Ephemeral architecture needs to be critical in the same measure as, say, the architecture of a museum: opening our being to death, while celebrating our human capacity to think the eternal. It doesn't do this merely by being technically responsible, built with light or recycled materials. For human life, even at its most precarious, seeks play and well-being, and never merely survival.

In my book *Polyphilo or the Dark Forest Revisited* (1992), I dealt with these issues in the form of a love story, a story of delayed material fulfilment that celebrates our human condition outside a simple dialectic, a condition characterized as 'bitter-sweet' by the Greek poets who invented Eros/Amor, the divinity of love, neither perpetually fulfilled nor perennially lustful. It is a story that celebrates the nomadic condition of modern technological man, caught in the liminal place of a fully carnal body and homogeneous mental space, always in transit, always crossing a threshold, travelling for the sake of the trip, rather than in view of a known destination. This narrative is also a theory of architecture as poetic image, suggesting alternatives to architectural practices based primarily on instrumental methodologies.

Instrumental theories have been dominant for two centuries. Most recent ones postulate the use of computers with a complete disregard for history and embodied consciousness (with its oriented spatiality). Leaving behind the computer's utilitarian justification as a tool that might improve the efficiency of architectural production, these theories claim the tool's capacity to generate 'new forms', totally 'other' from our traditional 'orthogonal' building practices. Indeed, recent powerful software packages are now capable of treating surface as the primary element in design, allowing for unimaginable configurations that are at once structurally sound and open up an infinity of formal possibilities. These instrumental processes are necessarily dependent on mathematical models, themselves designed by computer engineers working with specific economic interests in mind; extrapolated to architecture, they often become an empty exercise in formal manipulation. Fuelled by a host of technological dreams or nightmares, architects soon forget the importance of our spatial engagement (verticality) in this inescapable and particular form of bodily consciousness, with the world that defines our humanity, our historicity and gravity (the 'real world' of bodily experience into which we are born, and which includes our sensuous bond to the earth and all that is not human).

The reality of architecture is infinitely complex, both shifting with history and culture, yet also remaining the same, analogous to the human condition which demands that we continually address the same basic questions to come to terms with mortality and the possibility of transcendence opened up by language, while expecting diverse answers which are appropriate to specific times and places. Architecture possesses its own 'universe of discourse', and over the centuries has seemed capable of offering humanity far more than a technical solution to pragmatic necessity. Architecture and the design of the environment configure the 'limits' of human culture; they open a clearing for all the great achievements of art and civilization. Our technological world, obsessed by infinite progress and the obliteration of limits, is often sceptical about architecture having any meaning *at all* other than providing shelter. Yet our dreams are always set *in place*, and our understanding of others and ourselves could simply not *be* without architecture. We know architecture allows us to think and to imagine, it opens up the 'space of desire' that allows us to be 'at home' while remaining always 'incomplete' and open to our personal death, this being our most durable human characteristic. Without limits we simply cannot be human. Even cyberspace could not 'appear' if we were not first and foremost mortal, self-conscious bodies *already* engaged with the world through direction and gravity. We don't merely *have* a body, we *are* our bodies.

Architecture and design, operating at the limits of language, between nature and culture, between that which can be articulated in words and the unspoken and obscure, communicates to us the possibility of *recognizing* ourselves as complete, in order to dwell poetically on earth and thus be wholly human. The products of architecture have been manifold, ranging from the *daidala* of classical antiquity to the gnomons, *machinae* and buildings which Vitruvius names as the three manifestations of the discipline, from the gardens and ephemeral architecture of the Baroque period to the built and unbuilt 'architecture of resistance' of modernity such as Le Corbusier's La Tourette, Gaudi's Casa Batlo or Hejduk's ephemeral 'masques'. This *recognition* is not merely one of semantic equivalence; rather it occurs in experience, and as in a poem, its 'meaning' is inseparable from the experience of the poem itself. As an 'erotic' event, it overflows any reductive paraphrasing, overwhelms the spectator-participant, and has the capacity to change one's life. In order to propitiate such events the designer must necessarily engage language, a story capable of modulating intended actions (projects) in view of ethical imperatives, always specific to each task at hand. The practice that emerges with such a theory can never be an instrumental application, but rather appears as a 'verb', as a process that is never neutral and should be valorized. *This process in fact erodes the boundaries between the artistic disciplines concerned with space* and is not constrained by the specificity of the material or size of the designed object.

This has been the story of an architecture of resistance since Piranesi, passing through John Hejduk, Daniel Libeskind and Peter Greenaway. From the moment when the traditional divisions among the fine arts were subverted, first in epistemology and eventually in practice, between the eighteenth and the early twentieth centuries, the most significant works that 'construct' a mysterious depth, a significant spatiality, belong within my story. The book *Polyphilo* tells a story as a sequence of visits to some of these works, as a revelation for the man in transit, ephemeral truths that disclose and conceal simultaneously, maintaining a tension between the absence of gods and our desire and capacity to wait.

Polyphilo's plot is based on an older treatise, a most significant work in the European tradition of architectural theories, one that merits a short exposition and helps explain my own project. The original is entitled *Hypnerotomachia Poliphili* (Colonna, 1499), published in Venice in the late fifteenth century. The *Hypnerotomachia* is very different from other Italian Renaissance treatises on architecture with their rational emphasis and interest in mathematical principles, such as Alberti's *De re aedificatoria* or, later on, Palladio's *Quattro libri*. It is a Neoplatonic narrative articulating architectural meaning in relation to our human search for erotic fulfilment, a basic description of the human condition, equally applicable to the contemporary nomad. The interest of architecture is presented as a search for unity, an alchemical quest, through making. Architectural meaning is *experienced* as a sense of completion/order by the nomadic architect, in the mind's eye, before it is articulated in words or mathematical ratios, just as architecture is *made*, and its principles are 'found' rather than imposed from a concept, as a prescriptive recipe.

The story is a dream within a dream. Poliphilo dreams about being in a threatening dark forest and narrates the myriad things he saw, a veritable strife for love, which is the meaning of the Greek words in the title. In an erotic trance which is both fulfilment and the infra-thin space of perpetual expectation, he tells of many ancient marvels, architectural monuments, most of them in ruins, encountered in his search for Polia, his beloved. In this vein, always thirsty, he describes a pyramid and obelisks, a great horse, a magnificent elephant concealing the tombs of a king and a queen, a hollow colossus, and a triumphal gateway with its harmonic measurements and ornamentation (Figure 1.1). After suffering a major scare behind the threshold, passing the test of a frightening labyrinth, he is brought back to life by a wonderful encounter with five nymphs, embodying the five senses. They show him several fountains and he partially quenches his thirst by drinking the tepid water springing from a stone nymph's breasts. He is then taken to a magnificent bath where the nymphs disrobe and he is filled with desire, missing his Polia, architecture, always beyond the five senses. Eventually they arrive at the palace of a queen, who embodies free will. He is invited to a splendid meal and expresses his admiration for the variety of precious stones and materials that are worn by all those present, and describes a game in dance. After the festivities, he is taken to visit three gardens, the first made of glass, the second of silk, and the third containing a labyrinth. He eventually arrives at a crossroads: the site of three doors with mysterious inscriptions among which he must choose. Behind one Polia awaits him. Without either realizing the meaning of their physical proximity, she takes him to admire the four triumphs of Jove: four processions whose chariots and artefacts celebrate the stories of the classical poets explaining the effect of various kinds of

1.1 Woodcut from *Hypnerotomachia Poliphili* (1499) by Francesco Colonna

love. Then they witness the festival of Vertuno and Pomona and the ancient sacrifice of Priapus. Their journey seems to culminate with the description of a magnificent circular temple of great beauty, celebrating an *axis mundi* of water and fire, where the sacrifices of miraculous rites and ancient religion once took place. It is here that the couple fully acknowledge their loving encounter.

Yet, immediately after, Poliphilo and Polia are separated. The revelation brought about by place is but a brief event. Their human pilgrimage must proceed. They arrive at the coast to wait for Eros, at the site of a ruined temple where she persuades him to explore in search of admirable ancient things. He descends into a dark pit. There he finds, among many beautiful monuments and epitaphs of lovers separated by death, a mosaic mural depicting hell, a place without architecture, the place for men and women who have either loved excessively or have refused to love. Scared again, he returns to Polia, just in time to meet Cupid who has arrived in his ship propelled by beautiful rowing nymphs. Both climb aboard, and Love uses his wings as sails. Sea gods, goddesses and nymphs pay tribute to Cupid and the vessel arrives triumphantly at the island of Cytherea, across the sea of death. Poliphilo then describes the forests, gardens, fountains and rivers on the island, as well as the procession of triumphal chariots and nymphs in honour of Cupid. In the centre of the island, the final place of arrival, he describes the circular theatre made of alchemical glass with its venereal fountain and its precious columns. Mars makes love to Venus, followed by a visit to the innermost enclosure containing the tomb of Adonis, where the nymphs tell the story of the hero's death and of the sad celebration of his anniversary commemorated every year by Venus, the lover.

The nymphs then ask Polia to tell her own story, the origin and difficulties of her love. Polia agrees and her words fill the second book. She gives a genealogy of her family, explains her initial inclination to ignore Poliphilo, and provides a detailed account of the final success of their affair. Following Polia's account, Poliphilo concludes by describing their embrace, a passionate kiss, in the happy place of dwelling. Then he is awakened by the song of a nightingale and realizes that he has been dreaming. He is sad and alone, the space of love is the space of architecture, the space of the design arts. Neither exclusively path nor place, it is definitely a space with limits, and it is 'bitter-sweet', a place that may allow humanity to keep on dreaming, while *knowing* that it is dreaming.

Hypnerotomachia Poliphili's essential intention to articulate a possible ethical position through a narrative that acknowledges important models for the practice of architecture is very relevant. Creatively reading Heidegger, the Italian philosopher Gianni Vattimo has demonstrated how art works in our traditions may indeed embody 'weak truths', significant revelations for the present and suggestions for further action in the future, an alternative to the presumption of absolute truth in traditional religion, science and technology. Thus my modern rewriting of the Renaissance romance examines paradigmatic works of the twentieth century. In addition, Poliphilo's search, like ours, is never finished. The space of desire remains always open, yet limited. This is an architecture for a pilgrim seeking knowledge and initiation, always in transit and transformation, where even stone monuments acknowledge and celebrate their ephemerality. Also important is the basic phenomenological lesson of *Hypnerotomachia*, connecting architectural meaning to embodied experience through discourse, rather than simply accepting meaning as an effect of exclusively mental or linguistic processes.

In order to contemplate the potential architecture of the future as a *poiesis* or making that is not solipsistic, that is, capable of effective communication, phenomenology offers important insights. It leads to a recognition, through personal experience, of the body as the site of meaning, a body always engaged with a given world in a pre-reflective transaction, upon which other meanings are constructed. To this primary reality the work of art and architecture speak in the medium of the erotic. Erotic knowledge is never experienced by the mind alone. It always occurs in the world; it is of the flesh. It is knowledge of things by the individual body and is also of beauty, which is the whole. Erotic knowledge is a paradigm of 'truth as unveiling', the Heideggerian *aletheia*, as opposed to the objectifying 'truth as correspondence' inherited from Plato by Western science and metaphysics. It is a disclosure of *being*, one that is never given once and for all but that speaks of the presence of lighting, of the horizon that makes 'things' possible.

Plato himself, quite in opposition to his better-known articulation of truth, in the myth of the cave as a light that excludes all shadows, insisted that *agathon* (the sun) makes truth visible but is never to be beheld directly, for its force would blind us: truth and goodness are therefore not objects of pure contemplation, as in science, but are perceived through full bodily experience. The only time when humans may behold the sun directly is at dawn, a time that is, even today, the most propitious for the retrieval of poetic discourse, for speaking of the truth. This is the extended time/site of my own version of *Polyphilo*.

The modern hero also dreams, and the narration takes place entirely during crepuscular time. Polyphilo dreams about being in a threatening dark forest, the technological world with its capacity for human-generated apocalypse, and narrates the myriad things he saw, a veritable strife for the manifold love of all, which is the meaning of his name in the title. He tells of many postmodern marvels deserving of a place in the memory theatre of the future, portable architectural monuments encountered in his search for Polya, his beloved.

His dream takes the form of a trip through our modern, scientific geography, one that occupies three paradigmatic spaces: the private domestic space of the hero's bedroom, the public space of airports, and the space of the airplane, beyond a conventional categorization that now seems inoperative. These spaces are only modulations of the technological continuum, of a single quantitative universal entity, our 'intersubjective' reality, at odds with the mythical landscape of places that constituted the ground of classical or traditional architecture. The technological trip *par excellence*, air travel, occurs as a revolution of the globe, the totality of our finite space, at approximately 60 degrees North latitude during twenty-four hours of measured time. The technological trip is deconstructed through the paradigmatic human dream, the dream of flight, which underscores all architecture present and past still devoted to reconciling man, the vertical creature, with gravity. Polyphilo travels westward through the homogeneous space of our scientific universe; he is constantly in motion but arrives nowhere. And time is suspended, always the day's breaking lights. He is always in the same place yet visits magical architectural objects pregnant with meaning, works that constantly question the assumption of a universal, geometric space as the place of human existence. His time is therefore always the present, catching up with the time of his departure, the simultaneity of simulacra; yet it is always sunrise, a privileged time of day that propitiates the fictional articulation of human temporality and that may allow for a return of mythopoetic narratives, even now, as we find ourselves at the end of progressive history.

The modern dream of Polyphilo is a dream of flight, with its vertical motion, one still capable of going somewhere. This motion of the poetic imagination allows the modern hero to 'inhabit' the diverse works and 'deobjectify' them, describing their 'placemaking' characteristics, extracting a philosophical and ethical lesson for the architect of the future. These works range from known to unknown projects of designers, artists and architects, from figurative, metaphysical or surrealist to abstract painting, sculpture, literature and drawing, They are all portable.

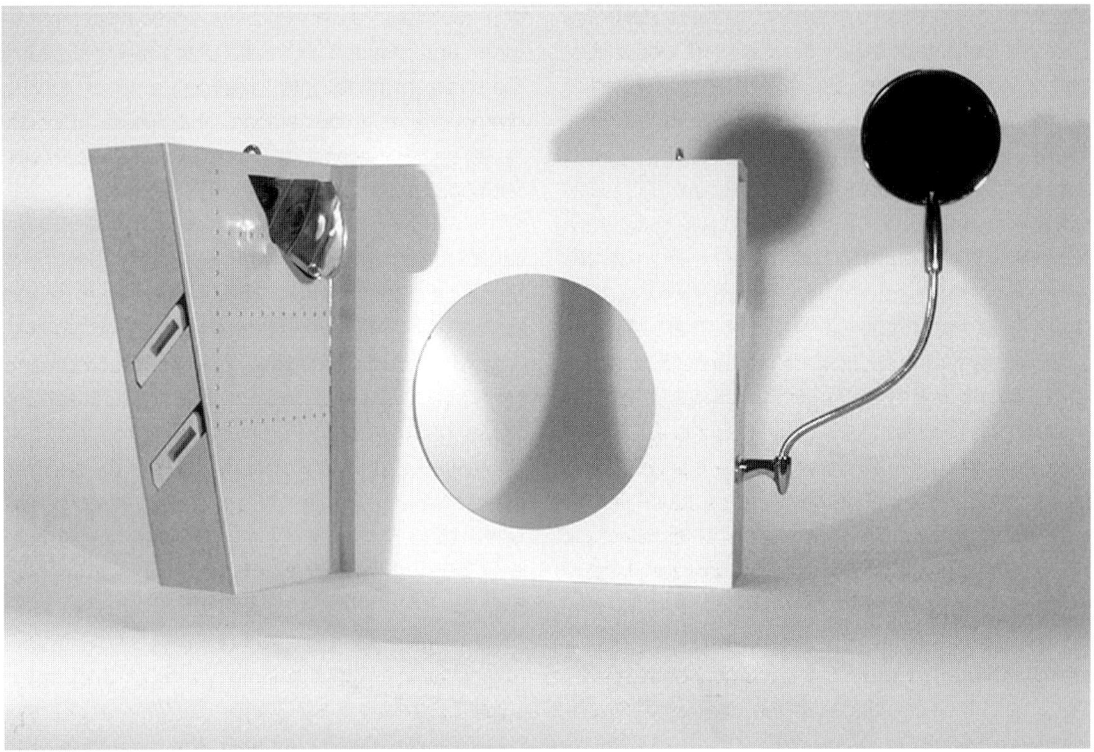

1.2 'Doctor' from the exhibition *Polyphilo's Thresholds* by Louise Pelletier, Martha Franco and Alberto Pérez-Gómez

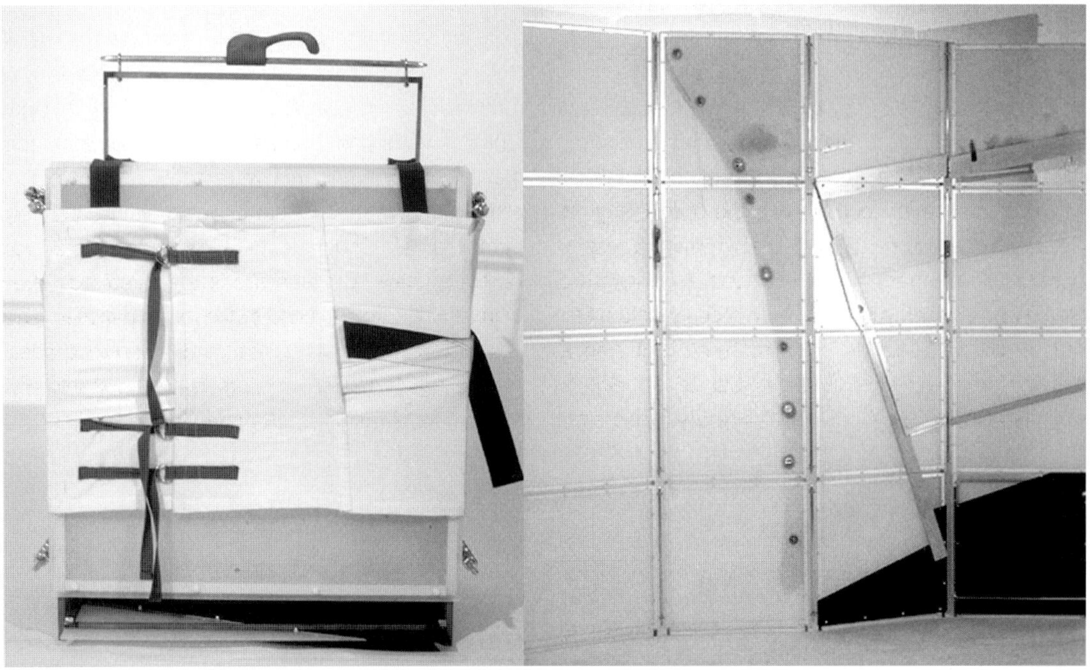

1.3 'Pantaleon' from *Polyphilo's Thresholds*

The text of Polyphilo has been the point of departure of many projects. Few of the selected works illustrated here include four portable objects intended to animate the anonymous space of transit, in this case, and with specific reference to *Polyphilo*, the waiting space of an airport. Drawing from the book and tacitly acknowledging the demise of 'place' (*genius loci*) as a culturally given condition, we recognized *threshold* was our operative context, stepping outside the traditional dialectic of path and place. The notion of threshold as a metaphor for architecture's liminal condition between reason and the irrational (at the limits/origins of language) is central, as it is also a fundamental concept in *Polyphilo*, where it is articulated as a typical condition of modernity since the end of the eighteenth century. The exploration of architecture/threshold is urgent in a world increasingly transformed by speed and new modes of communication. Threshold was given a specific programme by using 'commedia dell'arte' characters as a memory of the public realm, and 'objects/ types' of the domestic environment as a memory of the private.

These 'Polyphilic' objects, in the tradition of the Greek *daidala*, are constructed of well-adjusted and harmonious parts, and hopefully capable, as in the tradition of the *Iliad* and the myth of Daedalus, of transforming natural place into cultural space, revealing, through abstract construction, precision and metric wonder (eurhythmy), the 'luminosity' of the more-than-human world that makes humanity possible. In the early twenty-first century, these are objects for everyday use, yet marvellously unfamiliar, with a site in a domestic environment, but also a function in the perpetual threshold of an airport waiting room. These objects, originally generated from collages of recognizable aeronautical structures, are intended to 'make place' and transform the space of transit into 'dwelling', a space for the dramatic interaction among Pulichinella, Doctor, Angel and Pantaleon: an open plot between Man, Woman, Doctor and Angel; everyone's plot.

Doctor, a medicine cabinet (Figure 1.2), travels and becomes a 'personal care' case to use in an airport facility, particularly in case of emergency, to domesticate hygienic surroundings, perhaps allowing us to face ourselves in moments of solitude, or if the end arrives unexpectedly. Pulichinella, a light table, and Pantaleon, a dressing screen (Figure 1.3), become the centre and the skin of a nomadic dwelling, the perpetual dance of Hermes and Hestia, a repository of memories in the form of containers for images. And Angel's receptacle opens to become, only when utterly void, a *chaise-longue* on which to wait for angels, to listen for a light flapping of wings or for unintelligible whispers. Angel provides a sleeping bag for house guests, a dream site for a traveller in transit (Figure 1.4).

Only by acknowledging limits, and making architecture the site of such limits, can we be free and have access to real flight, to the dream of flight which reveals the human spirit (our breath) and that has nothing to do with infinite technological mobility. These objects are portable, yet deliberately cumbersome. They are difficult to assemble and relatively heavy, yet their site is the threshold. They animate the threshold and make the in-between alive with significance, beyond the traditional dichotomies. They are both rational and dream-like, portable and permanent, simultaneously uncanny and familiar.

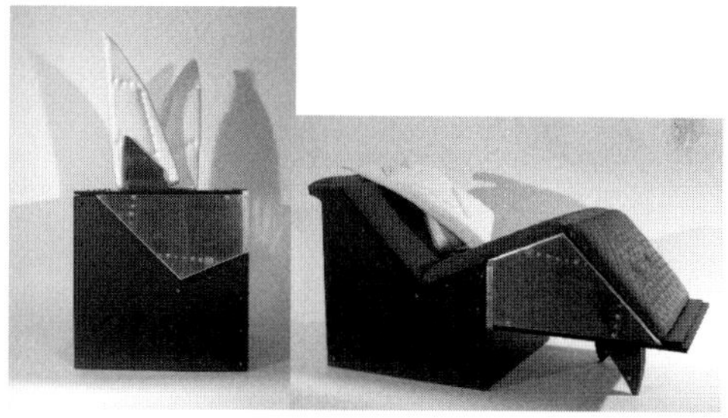

1.4 'Angel' from *Polyphilo's Thresholds*

References

Colonna, F. (1499) *Hypnerotomachia Poliphili*.
Pérez-Gómez, A. (1992) *Polyphilo, or the Dark Forest Revisited: An Erotic Epiphany of Architecture*. Cambridge, Mass.: MIT Press.

The Figure of the Spiral in Marcel Duchamp and Frederick Kiesler

Helmut Klassen
York University and Ryerson University

> [A]s you know, I am interested in the intellectual side, although I don't like the word "intellect." For me "intellect" is too dry a word, too inexpressive. I like the word "belief.". . . I believe that art is the only form of activity in which man as man shows himself to be a true individual. Only in art is he capable of going beyond the animal state, because art is an outlet towards *regions which are not ruled by time and space*. To live is to believe.
> (Duchamp, 1973, p. 137; emphasis added)

> I started to perceive space not as a void but as a link between every object, both of nature and man-made. There was a continuity held together by visually nothing, but this "nothing" was nothing else but the breath of the cosmos.
> (Kiesler, 1996, p. 134).

In this chapter, the modern experience of movement and mobility is situated within new theoretical and practical descriptions of space and the context of the urban milieu in the early twentieth century. It is proposed that movement and mobility in Modernist discourses of art, architecture, and technology represented more than the application of technological innovation to practical problems of social and economic efficiency. Movement and mobility also expressed utopian aspirations for new, unified expressions of space appropriate to the transformed conditions of modern urban experience. One pervasive motif of utopian spatial thinking in avant-garde art and architecture was the spatial and temporal figure of the spiral. The focus of my argument is on the idea that the figure of the spiral represents an ambivalent reconciliation of the experience of relentless change and movement found in the modern city with the promise of stable formal expression. I suggest that the spiral in avant-garde theory and practice produces a kinetic model of an indeterminate and primary spatial dimensionality as a foundational order of the city, supplanting the traditional idea of the city founded on the figure of the labyrinth. In the context of this chapter, I reconstruct the image and meaning of the spiral motif through an examination of the correspondences and exchanges between the artistic practice of Marcel Duchamp and the architecture of Frederick Kiesler from the 1910s to the mid-1940s.

In Duchamp's and Kiesler's evocation of the spiral figure, it is technology that frames the construction and sets in motion an image of spatial and temporal orientation within the chaotic existence of the modern city. The rhythms of the city – both the physical movements produced by technological innovation but also the social practices required to keep technological systems in motion – had an ambivalent reception in the nineteenth and early twentieth centuries. Paradoxically, new scales of productive activity resulted in increased disruption to people's lives, including constant destruction of the existing city precisely in order to rebuild it in a new image. A consequence of destruction was the dislocation of communities and the impossibility of forming memories of place in relation to fixed landmarks and repetitions in everyday life.

Nevertheless, the new technological conditions of production and mobility also produced new forms of experience. The early actuality film by the Edison Manufacturing Co., *A Trip from New Brooklyn to New York via the Brooklyn Bridge* of 1899, presents a unique image of modern mobility produced by the technology – the train – that in other respects had so disrupted the form and life of the city. The camera, placed on the front of a train traveling from Brooklyn to Manhattan over the Brooklyn Bridge, presents the film viewer with an almost magical experience of transformation. At the center of the bridge, the viewer of the film experiences a brief moment in which the everyday spatial context of city is lost in a virtual tunnel produced by the movement of the camera in relation to the rhythmic structural members of the bridge. The experience is that of continuous movement into a space whose depth appears to recede indefinitely (Figure 2.1). This image of movement, produced technologically by the moving train, the camera, and the system of cinematic projection, concentrates the experience of movement into a mesmeric spatial form. The framing of the camera – from a viewpoint beyond the particular experience of any one individual – becomes a paradoxical means of domesticating the incessant, chaotic experience of movement in the city.

I suggest that mediation by the technologies of physical movement and their transformation through the production of a mass cinematic image perform a task that is often lost in discussions of movement and mobility. These images transformed the abstract, spatial extension of modernity by reintroducing spatial directionality, orientation, and meaning for a mass subject. Movement here becomes more than an abstract state of being (or loss of being); through the mediation of the cinematic image, and through art and space in Duchamp and Kiesler, movement and mobility are grounded in a new, kinetic experience of space, more so than being the physical overcoming of spatial limits.

In the experimental work of Duchamp and Kiesler, a new description of space may be summarized by the idea of kinetic motion, both virtual and actual. The kinetically charged image signifies a space of social and metaphysical freedom for Duchamp and the reconciliation of man with nature and art for Kiesler. By examining selected projects of each artist related to the figure of the spiral as well as spatial and architectural designs in which they were both involved between the years 1915 and 1945 – focusing on their conceptual correspondences but also the divergences in their intentions and realizations of different environments – I articulate the meanings and significance of movement and mobility in this era and broadly discuss some of the implications for the theory and design of transportable environments.

Duchamp and Kiesler: points of correlation

Kiesler's essay on Marcel Duchamp's *Large Glass*, entitled "Design – Correlation: Marcel Duchamp's Big Glass" (Kiesler, 1996), suggests that he saw in Duchamp's masterwork a confirmation of his own theory of *correalism* in which he emphasized a new unity between man, nature, and art. Kiesler's interpretation of the *Large Glass* in this article was idiosyncratic. Presumably guided by his work as an architect, it passes over other interpretations of the *Glass* as a modern allegory of desire to emphasize the *Glass*'s Constructivism. He states that painting, sculpture, and architecture have become one in the *Glass* (Kiesler, 1996, p. 38). He sees Duchamp's use of plate glass as similar to the architect's use of glass in modern building. For Kiesler, plate glass is the paradigm of modern materiality as it has the unique ability to both divide and connect space. Further, the traditional expression of painting and sculpture is transformed in the *Large Glass* into the direct expression of "naked" materials technologically handled and poetically applied. Direct expression of materials in place of traditional ornamentation of surfaces is

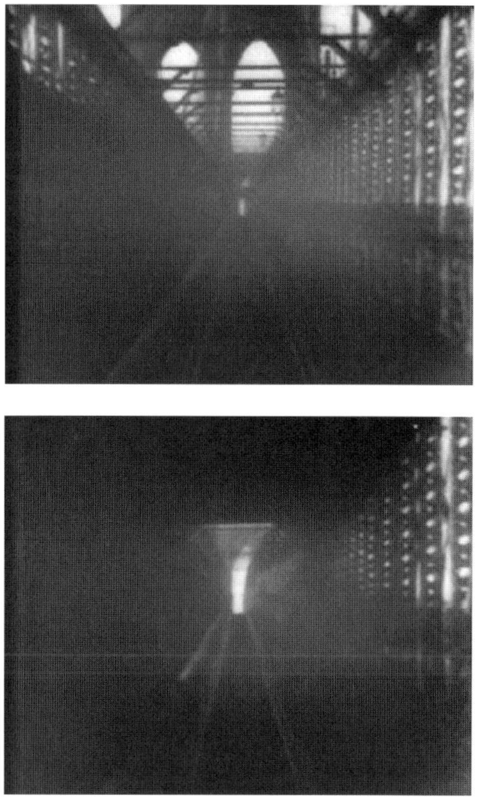

2.1 Thomas Edison. *Film stills: A Trip from New Brooklyn to New York via the Brooklyn Bridge*, 1899

also a fundamental paradigm of modern design (Figure 2.2).

Ironically, part of Duchamp's artistic persona was his renunciation of the role of artist along with a critique of theories of individual artistic expression. As part of this critique, he actively encouraged the interpretation of his work as anesthetic, the indifferent product of engineering and scientific and mathematical research more than the expression of any personal artistic vision (Henderson, 1998). Among Duchamp's works, many are constructed interventions that confound normal descriptions of space. *Sculpture for Traveling* of 1917 was a portable object consisting of brightly colored rubber bathing caps cut up into strips and stretched across his studio on string. The better-known *Door* was a reconstruction of the material fabric of his studio such that the door was paradoxically always both open and closed (Figure 2.3).

The majority of Frederick Kiesler's architectural work was in the areas of theater, shop display, and exhibitions. Though he designed many buildings, with the notable exception of the *Shrine of the Book* in Jerusalem they remained unbuilt. Kiesler was part of or acquainted with many of the avant-garde movements in the early twentieth century: the Vienna Secession, Russian Constructivism, De Stijl and Surrealism. Common to each of these movements was an emphasis on the construction of a new unity of space appropriate to modern life. It is in the context of the idealist aspirations of these movements that the figure of the spiral emerged in his work in the theater. In 1924, he built a spiral stage construction in Vienna known as the *Space Stage* that demonstrated new plastic ideas promoting direct contact and continuity between performance and audience in an all-round setting (Figure 2.4). He subsequently designed a prototype for an *Endless Theatre* characterized by a circular plan and egg-shaped section. In this design, the figure of a double spiral ramp formed the stage and spiral ramps mediated the plan organization and complex levels of audience seating. The fundamental idea guiding the precise geometrical articulation of the theater was the intertwining of the implied movement of the audience and performers in relation to the spiral figure as well as the proposed physical movement of various elements of theatrical mechanisms, stage, and seating to adjust for various types of performances.

Over the years and in many types of projects including theatrical architecture, commercial design and display, exhibitions, and housing, Kiesler continued to innovate with the integration of spatial design and the incorporation of various forms of virtual and actual moving building elements to advance his concept of a flexible kinetic space appropriate to the manifold functionality he theorized as critical to the modern age.

Duchamp's *Anemic Cinema*: indeterminate space

From the early 1920s through the 1930s, Duchamp produced a series of optical disks, later known as *Rotoreliefs*, patterned on the model of his *Bicycle Wheel* readymade. These consisted of circular disks with patterns based on variations of the spiral motif. Once set in motion (on an apparatus that Duchamp had designed, built and tried to sell at an inventor's fair as a novel, technological experience) the patterns produced the optical illusion of an ever-receding spatial depth or, depending on the patterns, spatial projection (Figure 2.5). In conjunction with Man Ray, in 1926 Duchamp created a short film based on the idea of the optical disks entitled *Anemic Cinema*. In addition to various spiral patterns, he interjected an explicit sexual content through alliterative phrases mimicking the arrangement and movement of the spiral pattern in between the presentation of each different spiral design. Of the many meanings evoked in the film, it is the nature of the kinetically produced image of space that bears on this chapter.

2.2 Marcel Duchamp. *The Large Glass*, 1915–23

2.3 Marcel Duchamp. *Door, 11 rue Larrey, Paris*, 1927

2.4 Frederick Kiesler. *Space Stage*, Vienna, 1924

2.5 Marcel Duchamp. *Rotary Demisphere* – Precision Optics, 1924

I would suggest that the spatial image produced through Duchamp's apparatus intentionally thwarts our habitual perception of objective space and time. It reveals a space in which all objective determinations, including that of the self, are rendered indeterminate through the principle of conceptual and actual kinetic movement.

Much speculation upon Duchamp's work, especially the *Large Glass*, centers upon ideas of a fourth dimension of space. While the significance of the fourth dimension in Duchamp's work has been debated, it remains significant that his understanding of advanced mathematics and geometry far exceeded a cursory interest, attaining a level that was the most advanced of any artist of the time (Adcock, 1983). Speculation on the fourth dimension traversed mathematical, scientific, artistic, and spiritualist discourses of the late nineteenth and early twentieth centuries. Mathematically, the idea of the fourth dimension derived from the development, in the early nineteenth century, of various forms of non-Euclidian geometry. It is significant that these new geometric systems severed the link between the logic of mathematics and the experience of the world apprehended by the senses. Constructed speculatively on the questioning of Euclid's fifth postulate, to be true these geometries did not need verification in the world but primarily required internal, logical consistency. The result was that worlds other than the one perceived by the senses could be postulated conceptually. The possibility of a fourth and, further, *n* dimensions, in science and art in the early twentieth century was the result of this epistemological shift.

Duchamp's artistic production rests on the paradoxes produced between conceptual and perceptual knowledge and expression. It is impossible for four- or more dimensional *objects* to be directly apprehended by the senses as their hypothesized four-dimensional reality always results in an impossible three-dimensional reality. To be fully perceived they would have to be seen from many different points of view at the same time. Nevertheless, four-dimensional objects could be evoked to our understanding through approximations of their character. The critical link in imagining space and time from the third through to the fourth dimension was movement. Adcock (1983) has stated that Duchamp's readymade *Bicycle Wheel* is a conceptual approximation of the moving reality of the fourth dimension through its potential double movement, simultaneously around its own axis of rotation as well as around the axis of the forks. *Anemic Cinema*, as a moving image, puts an accent on the kinetic production of an approximation – the illusion – of an anesthetic, conceptual space outside of the habitual experience of the three-dimensional world.

What is seen in the readymades, *Anemic Cinema*, and the *Rotoreliefs* is an expression of movement different from but nevertheless consistent with his earlier, famous painting *Nude Descending a Staircase (No. 2)*. The difference resides in the anesthetic quality of Duchamp's work already discussed. The earlier abstract painting refers to the representation of human movement abstracted through stop-motion photography developed by Marey and Muybridge. In contrast to the logic of abstraction, the rotating optical disks cannot be said to represent anything. Even the rotating textual segments – alliterative phrases with overt sexual connotations – have been described as an attempt to transform language into kinetic, visual hieroglyphics rather than to express logical discursive meanings (North, 2002). The rotating disks produce an illusion from the relation of kinetic rhythms between the technical apparatus of the rotating spiral, the direction of the movement in the film, and the body of the viewer.

As Judavitz (1996) has noted in her analysis of *Anemic Cinema*, the most significant (among many) allusions of the moving spirals is that of an eye. With this allusion, the conceptual spatiality of the fourth dimension produced in the film can also be understood as being projected out of the screen into the space of the viewer. That is, the eye of the film screen *looks* back onto the viewer and encompasses him/her into the kinetic space of the film. This suggestion is strengthened by Duchamp's intention to project the film onto a translucent screen backed by a mirror coating. Further, Adcock (1983) notes that mirror images are one of the tropes of the fourth dimension in our experience of the world. This is because the geometrical translation of an object from three to four dimensions and back again (*round-trip*) produces an inversion of the original object similar to that of the image in a mirror. In *Anemic Cinema*, this is suggested by the title frame that graphically produces the words *anemic* and *cinema* as an anagrammatic mirroring (Figure 2.6).

In the context of the fourth dimension, *things* in Duchamp's work are no longer defined by objectivity but by an oscillating movement between visible and invisible states of being and meaning. Through the projected experience of the four-dimensional space, the viewer is incorporated into an indeterminate dimensionality that remains in a constant state of physical and conceptual transformation. The fixed boundaries of things – including consciousness of self as distinct from that of the world – are transgressed from the outside by kinetic operations. While Duchamp rejected any direct spiritual meaning of the fourth dimension, whether of traditional religion or its spiritualist shadows, the hypnotic or mesmerizing effect of the moving spirals nevertheless refers to a radical state of actuality, the dissolution of consciousness, and the collapse of the limits between subject and object, a condition that was discussed as a spiritual practice of attaining higher consciousness in late nineteenth-century spiritualist discourses.

Kiesler: the city, the shop window, and desire

Kiesler's book entitled *Contemporary Art Applied to the Store and its Display* (1930) maintains his early concerns for a new unity of man, nature, and art. Significant for the discussion in this chapter is how Kiesler implicitly frames the ideal of a new spatial reality represented by his earlier work on the *Space Stage* and *Endless Theatre* in terms of the workings of human desire through the ubiquitous shop window of retail stores. I would suggest that Kiesler's interest in and unique interpretation of Duchamp's *Large Glass* discussed above resonated with the ideas expressed in his earlier book. Though his text is specific to the problem of promoting sales of luxury consumer goods, we may also read into this work an analysis of the alienation of human social relations in the modern city and a spatial model capable of countering this alienation.

Kiesler states that the plate-glass shop window is the plane of contact between the consumer (significantly, female) and goods for sale. From the sense of touch familiar in the traditional market place – the ability to handle the goods directly – there was a change to an appropriation of things by the eye, a condition that extended beyond the situation of the shop window to include all of modern culture. Kiesler states that the glass shop window is a specific medium of communication between the public and the store that requires its own form of design response to ensure that the possibility of contact enabled by the glass *screen* is in fact made and the sale of goods consummated. His critique of store display techniques is that they have not understood the perceptual transformation occasioned by the shop window, treating the window like a storehouse of goods in the manner of traditional bazaars. In the context of the new visual regime of modernity, this approach overwhelms any advantage that the glass window provides. Without proper perceptual research, creative production, and a unified design effort to adjust the approach to the display and viewing of goods, the plate-glass window is as much an alienating barrier to contact with goods as an opaque wall would be. Implicit in his commentary on the shop window is the idea that life in the modern

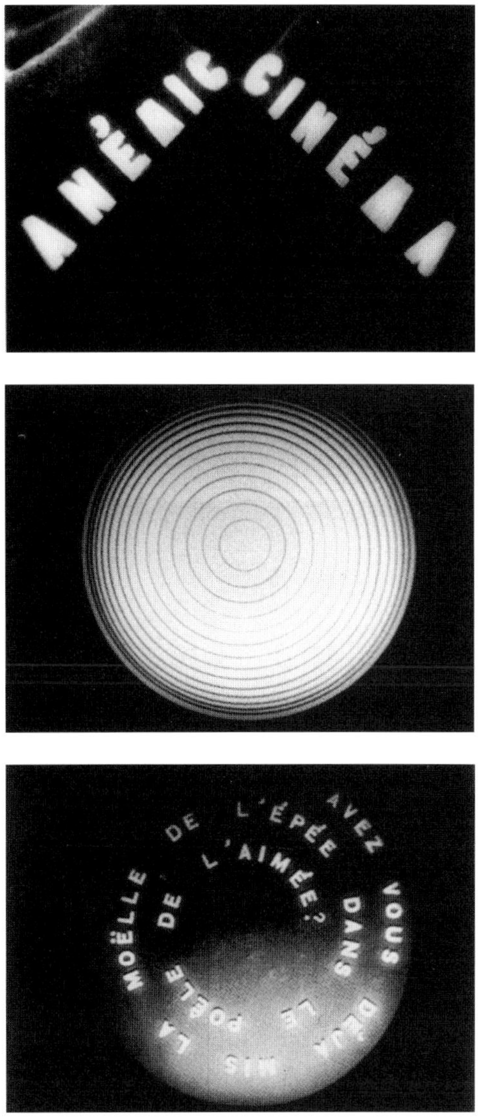

2.6 Marcel Duchamp. *Anemic Cinema*, 1926 – Title Frame, Spiral, and Pun

2.7 Frederick Kiesler. *Jay's Shoes*, Buffalo, NY, 1935

city is characterized by estrangement. Each person is an isolated individual following his or her own determined path because there are no inherent forces in the structure of modern human relations to bring people together. Proximity and chance confluence of individual paths cannot create the conditions of social contact. As a response to this condition, Kiesler states that the medium of the window and the message of the display of goods must work to *stimulate desire* in the passerby, actively attracting *her* to the window and subsequently into the store.

The solution to stimulating desire for Kiesler can be described as primarily spatial in character, concerned with animating the *empty* space that encompassed the situation of the object of desire and that of passersby. What Kiesler proposes is reminiscent of the effect of Duchamp's *Anemic Cinema*: the store front should be designed as a *funnel* that creates a *suction-like effect* on the passersby, pulling them off their trajectory on the sidewalk toward the merchandise in the window and the entrance to the store (Figure 2.7). Oriented first to the eye and then to the body of the passerby, the shop window becomes a spatial mechanism, reminiscent of the cinematic tunnel in Edison's film and Duchamp's *Anemic Cinema* that incorporates the passerby into *its* effect. After diagrammatically illustrating how the spatial funnel is functional in terms of drawing people into the store and efficient in terms of length of display created, Kiesler generalizes the commercial model to the theater, cinema, exhibitions, and other architectural designs. His text describes a new kind of architectural object modeled in three dimensions to present a *full plastic state*, an idea echoed in countless other theoretical expressions as an ideal of modern architecture in contrast to low-relief or decorative elaboration of surfaces. Recalling his earlier *Space Stage* and *Endless Theatre*, I would suggest that his evocation of a three dimensional plastic state, one in which the object is seen from many different angles and at different times, in fact referred to the development of spatial tactility in which an indifferent relationship between people and things would become erotically charged. His ideal was of a continuous architectural space that was actively oriented – correlated – to the physical, perceptual, and emotional appropriation by the subjects of the built environment. That is, the built architectural object is in erotic continuity with the space that it shapes, both building and space genetically formed not on functions abstracted into universal standards of design but on the idea of function conceived elastically as a problem of how one should live in the present (mechanical) age as a complete human being.

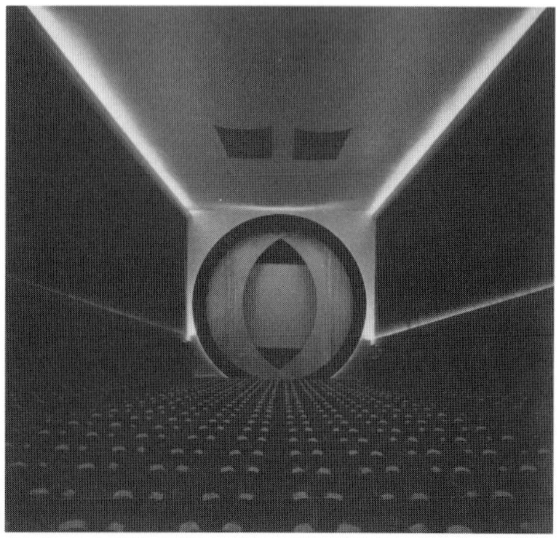

2.8 Frederick Kiesler. *Guild Cinema*, New York, 1929

As has been noted in Daniels (1996), the *Guild Cinema* design of 1929 that Kiesler illustrates near the end of his book shows a conceptual and physical space in which one can imagine Duchamp's indeterminate illusion of space projected in *Anemic Cinema* capable of being actualized (Figure 2.8). Among many of its innovative design features, the flexible screen of projection was conceived as an eye/camera projecting the space of the film back onto the audience.

In his book, Kiesler describes the co-relational effect of space in architecture as *psycho-function*, a term by which he meant the "surplus above efficiency which may turn a functional solution into art." Art for Kiesler signified the "the organization of static materials into a living unity." Artistic unity was not characterized by static relationships but by the *principle of flux* found in the creations of nature and in art forms such as music and the technologically mediated form of cinema. Architecture was, therefore, an artistic expression of continuity and unity characterized by a rhythmic articulation of flux and movement. In the context of the modern city, architecture served a redemptive function in recognizing and developing its kinetic potential to stimulate desire and facilitate contact and exchange among people, things, and places out of inherently estranged flows and indifferent spatial organization. The architect achieves the ideal of kinetic exchange by using all spatial and technological means available in the service of what Kiesler terms *creative production*, a concept that he opposes to the overwhelming predominance of *reproduction* in modern industry. Thus, for the shop window, Kiesler articulates a program in which the organization of the spatial funnel, already discussed, is correlated in rhythmic relation to the design of multiple and contrasting lighting elements and projected images, focused not only on the merchandise but also innovatively integrated into the elements of sidewalk, floor, ceiling, advertising, and the interior of the store itself. He further articulates the idea of future technological innovations that would render the shop window fully kinetic: a button-controlled robotic apparatus by which passersby would be able to handle and view the merchandise they desired behind the glass; and the development of a form of television that would render the exchange through the shop window interactive. Nevertheless, in his vision of a kinetic architecture, technology is never autonomous but is always mediated within the *endless* spatial ideal.

Surrealist exhibition installations

From a series of Surrealist exhibitions and window installations realized in the 1940s, I will briefly examine two exhibitions where Duchamp and Kiesler respectively had a major role in the construction of the display environment: *First Papers of Surrealism* and *Art of this Century*, both in 1942. Of all the possible works by Marcel Duchamp and Frederick Kiesler, the Surrealist exhibitions are the field of operation in which the ideas of these two figures of the avant-garde historically converged and at the same time contrasted with each other. As discussed above, Duchamp's 1926 film *Anemic Cinema* articulated the description of an indeterminate space in terms of the problematic of human vision. This was foreshadowed in the *Large Glass*, a work that was very influential on Kiesler at least from the mid-1930s. Duchamp continued working on the *Rotoreliefs* into the late 1930s. While Kiesler had earlier articulated the problem of the shop window display in terms of a priority on human vision in modern social relations and artistic production, it was after his work on the theoretical articulation of shop window display that he began a series of explicit investigations into the mediation of vision through the design of display environments known as *Vision Machines*. It is the fruits of these researches by Duchamp and Kiesler that present themselves in the *First Papers of Surrealism* and *Art of this Century* exhibitions.

2.9 Marcel Duchamp. *Mile of String*, *First Papers of Surrealism*, New York, 1942

Organized by André Breton in 1942, the *First Papers of Surrealism* exhibition was the first gathering of Surrealist work in New York by the group of artists recently exiled from Europe. Housed in a classically designed mansion, Breton entrusted Duchamp with transforming the exhibition space for the paintings into something more appropriate to the Surrealists' revolutionary artistic program. His solution was the installation of a reputed one mile of string (Figure 2.9). The effect of the string was to foreground the space of the exhibition itself as an explicit part of the display, consciously engaging the viewers in their movement through space to look at the paintings. It simultaneously disturbed the possibility of any transparent act of looking. Resonating with his *Traveling Sculpture* of 1917, Duchamp's intervention rendered the exhibition space ambivalent: the gesture frustrated viewers' expectations of looking at paintings. Nevertheless, the gesture also had the effect of activating the space in a way that expelled the idea of the space as an empty, neutral container. Space was transformed into more than *no-thing*.

The *Art of this Century* exhibition of the same year had considerably more resources applied to its realization. Peggy Guggenheim engaged Frederick Kiesler to design and build the exhibition to display a wide variety of modern art. In the Surrealist Gallery, we can see the realization and development of many of the ideas that Kiesler outlined in his book on the store and display (Figure 2.10).

2.10 Frederick Kiesler. Surrealist Gallery, *Art of This Century*, New York, 1942

The primary characteristic of the space is that of a spatial mechanism reminiscent of the spatial funnel effect of the shop window and Duchamp's *Anemic Cinema*. The effect of the funnel is emphasized by the perception of a *floating* space of indeterminate depth created by the curved walls and the frameless mounting of the artworks. The virtual kinetic perception of the space is supplemented by the actual kinetic potential of each art work to be directly handled and moved by the viewer that recalls Kiesler's idea of a fully kinetic shop window. Finally, the lighting of the space was timed to illuminate only one side of the gallery at a time, alternating from side to side precisely every few seconds. This set up an effect of rotation that rhythmically enhanced the perception of the spiral configuration in the space as a whole.

In the Kinetic Gallery, Kiesler developed various versions of his ideas for *vision machines* to display different artists' work. The display of Marcel Duchamp's various works in the form of his *Box in a Valise* was mediated through a *peephole* device activated by rotating a large wheel on which the spokes were tied together, significantly, by a spiral band (Figure 2.11). In each case where the idea of *vision machines* is employed to mediate the viewing of art works, and in the attitude to the art pieces in the exhibition as a whole, Kiesler ensures that the viewers themselves are active in handling the displayed work. Whether works are displayed in an open gallery or through the mediation of vision machines, I would suggest that Kiesler's design intentions nevertheless follow up on the problematic he articulated in relation to the shop window; that is, the stimulation of desire through mechanisms of spatial mediation that focused the subject's attention physically and perceptually across the distance that separates them from the work, however great or small that distance may be.

Conclusion

Distinctly modern in the articulation of the figure of the spiral in spatial design is the construction of a viewpoint through which the body of the subject is immersed within the environment in the same way as the viewing audience is presented the projected film in the cinema. The result is that the experience of space and the city becomes unavoidably mediated through a technical apparatus. A potential consequence of this condition is a further distancing and estrangement of knowledge from experience through the construction of abstract points of view above or outside of the body and perception. However, both Duchamp and Kiesler consciously avoid this possibility, redeeming the unavoidable and apparently transparent position of modern mobility by spatially incorporating our eyes *and* our bodies into the effects of the technological apparatus.

Nevertheless, it is an ambivalent project. In Duchamp's anesthetic work we are promised an experience of engagement with an activated space of the world. Yet it is an encounter in which reconciliation is an experience endlessly delayed. In contrast, Kiesler appears in his work to have accepted the transformed conditions of movement and perception. Further, he used these conditions as the creative basis from which to explore the potential of art to reconcile human experience with the technologically driven world of the city. From the contrasting perspectives of indifference and delay to an effort to reconstruct a form of ritual participation in architecture – from anesthetics to esthetics – Duchamp and Kiesler remind us that both positions in relation to the transformed conditions of mobility in the modern city have nevertheless been extremely productive of new forms of kinetic perception and experience. They remain for us complementary pointers to address current theoretical issues in the field of transportable environments.

2.11 Frederick Kiesler. Paternoster for viewing Duchamp's *Box in a Valise*, Kinetic Gallery, *Art of this Century*, New York, 1942

References

Adcock, C. (1983) *Marcel Duchamp's Notes from the Large Glass: an N-Dimensional Analysis*. Ann Arbor, Mich.: UMI Research Press.

Adcock, C. (1989) "Duchamp's Eroticism: A Mathematical Analysis" in R. F. N. E. Kuenzli (ed.), *Marcel Duchamp: Artist of the Century* (pp. 149–167). Cambridge, Mass.: The MIT Press.

Béret, C. (ed.) (1996) *Frederick Kiesler: Artiste-Architecte*. Paris: Editions du Centre Georges Pompidou.

Bogner, D. (1988) *Frederick Kiesler: Architekt, Maler, Bildhauer, 1890–1965*. Vienna: Locker.

Bogner, D. (1997) *Frederick Kiesler: Inside the Endless House*. Vienna: Bohlau.

Daniels, D. (1996) "Points d'Interférence entre Frederick Kiesler et Marcel Duchamp" in C. Béret (ed.), *Frederick Kiesler: Artiste-Architecte* (pp. 119–132). Paris: Editions Centre Georges Pompidou.

Duchamp, M. (1973) *The Writings of Marcel Duchamp*. New York: Da Capo Press Inc.

Edison, Thomas (1899) *New Brooklyn to New York via Brooklyn Bridge, no. 2*. New York: Edison Manufacturing Co. Retrieved April 28, 2004, from http://memory.loc.gov/cgi-bin/query/D?papr:1:./temp/~ammem_hkjr::.

Gough-Cooper, J. A. J. C. (1989) "Frederick Kiesler and the Bride Stripped Bare . . ." in Y. Safran (ed.), *Frederick Kiesler, 1890–1965* (pp. 62–71). London: Architectural Association.

Held, R. L. (1982) *Endless Innovations: Frederick Kiesler's Theory and Scenic Design*. Ann Arbor, Mich.: UMI Research Press.

Henderson, L. D. (1998) *Duchamp in Context: Science and Technology in the Large Glass and Related Works*. Princeton, NJ: Princeton University Press.

Judavitz, D. (1989) "Rendezvous with Marcel Duchamp: Given" in R. F. N. E. Kuenzli (ed.), *Marcel Duchamp: Artist of the Century* (pp. 184–202). Cambridge, Mass.: The MIT Press.

Judavitz, D. (1996) "Anemic Vision in Duchamp: Cinema as Readymade" in R. Kuenzli (ed.), *Dada and Surrealist Film* (pp. 46–57). Cambridge, Mass.: The MIT Press.

Kachur, L. (2001) *Displaying the Marvelous*. Cambridge, Mass.: The MIT Press.

Kiesler, F. (1930) *Contemporary Art Applied to the Store and its Display*. New York: Brentano's.

Kiesler, F. (1966) *Inside the Endless House: Art, People, and Architecture: A Journal*. New York: Simon and Schuster.

Kiesler, F. (1996) *Frederick J. Kiesler: Selected Writings*. Stuttgart: Verlag Gerd Hatje.

Kiesler, F. (2001) "Manifesto on Correalism" in *Frederick J. Kiesler: Endless Space* (pp. 91–95). Ostfildern-Ruit: Hatje Cantz.

North, M. (2002) "Words in Motion: The Movies, the Readies, and the Revolution of the Word" in *Modernism/Modernity*, 9(2), pp. 205–223.

Otwell, A. (1997) *View*, Marcel Duchamp special issue, 1945: Chapter 3 "'Les Larves d'Imagie d'Henri Robert Marcel Duchamp': the Kiesler triptych." http://www.heyotwell.com/work/arthistory/thesis/chapter3.html. Retrieved December 20, 2003, from the World Wide Web.

Otwell, A. (1997) *View,* Marcel Duchamp special issue, 1945: Chapter 4 "Kiesler's theories and the Large Glass.' http://www.heyotwell.com/work/arthistory/thesis/chapter4.html. Retrieved December 20, 2003, from the World Wide Web.

Phillips, L. (ed.) (1989) *Frederick Kiesler*. New York: Whitney Museum of American Art in association with W. W. Norton.

Safran, Y. (1996) "L'Angle de l'Oeil, La Vision Machine de Frederick Kiesler" in C. Béret (ed.), *Frederick Kiesler: Artiste-Architecte* (pp. 133–141). Paris: Editions Centre Georges Pompidou.

Zentrum Wien, M. f. M. K. F. M. (2002) *Frederick Kiesler: Art of This Century*. Ostfildern and New York: Hatje Cantz, D.A.P.

10,000 Songs in Your Pocket: The iPod® as a Transportable Environment

Prasad Boradkar
Arizona State University

Introduction: object as environment

Transportable environments are typically understood as portable structures that can be relocated with relative ease, erected and dismantled quickly, constructed from non-traditional materials or deployed as short-term shelters. However, this merely represents the physical dimensions of environments created by architectural enclosures. Environments may also be conceived as mental rather than corporeal, imagined rather than built, ethereal rather than corporeal, and perceived rather than prototyped. Also, such environments may be generated by portable gadgets. Thrown into handbags, slipped into pockets, or simply hand-held, devices such as mobile telephones, MP3 players, and personal digital assistants allow us to create transportable environments that we can travel with. They contain fragments of other spaces – places we are temporarily away from. The object of this chapter is to study one such product, Apple Computer's iPod (see Figure 3.1), which through personalized music gives us the ability to transport our cherished environments with us, no matter where we might be. The environment created by the iPod is an assemblage of body, technology, and space. It is auditory and it contains treasured sounds; it is virtual and contains communities; it is ethereal and contains sensations.

Scholars in cultural studies have played a major role in decoding the processes by which people make sense of their private presence in the social world. One may acquire a better comprehension of the relationships between the individual and society, music and technology, and the body and space by referencing theories from this field of study. And, when accompanied by observations of lived experiences with objects, they can provide a means by which to decipher our environment. As a part of the primary research for this article, iPod users were observed and interviewed with the principal aim of finding out how these objects become a part of their daily routine and how they influence their relationship to the environment. Member websites that function as forums for the discussion and exchange of 'iPodding' information were accessed, along with weblogs where daily minutiae of iPod-related activities are recorded.

Mobile miniatures

> It is now a commonplace observation (to the point of weary cliché) that the explosive combination of tiny, inexpensive electronic devices, increasingly ubiquitous digital networking, and the world's rapidly growing stock of digital information is dramatically changing our daily lives.
> (Mitchell, 2002, p. 50)

Following this forthright statement in his essay "E-Bodies, E-Buildings, E-Cities," Bill Mitchell asks what this condition suggests for design in the twenty-first century. Since the arrival of digital technology in the early 1980s, this question has been raised numerous times by many scholars and there have been several rejoinders. There have been calls for the end of design as well as for the birth of a new design, requests for a more socially informed design, references to a second modernity, and so

3.1 The iPod® (Courtesy of Apple Computer)

on. Common observations, however, are of a definite and steady miniaturization and dematerialization (also referred to as immaterialization or demassification) of objects and shorter design cycles due to quicker communications. Digital circuitry has replaced electromechanical machinery, integrated chips and printed circuit boards have triumphed over motors and pulleys, and moving electrons have deposed moving components. The result has been an explosion in the number of small, digital gadgets like laptop computers, personal digital assistants, mobile telephones, MP3 players, and video games. These products that have made us more mobile, more connected, and more free from cumbersome wires plugged into walls have become almost indispensable in the daily functioning of work as well as leisure activities in today's urban life.

Digital music cultures

In the case of music, the advent of digital technologies triggered the shift from analog to digital signal and had a significant impact on patterns of production and distribution. However, consumption of popular music changed with the stellar rise of a new file format. Called MP3 (which cumbersomely expands to Motion Picture Export Group-1 Audio Layer 3), these audio files are much smaller in size, and in consequence immensely portable. The quick proliferation of these files was aided by a growing number of high-speed Internet or broadband connections in households, dormitory rooms, computer laboratories at universities and business locations, as well as by cheaper and better computing equipment.

> From the time that computer makers began including CD drives in personal computers (augmented later with good quality audio speakers) the music listening habits of many white-collar workers, students and others who use personal computers on a daily basis have undergone significant change.
> (Jones, 2000, p. 218)

The MP3 file is as versatile as it is portable, because it can be "ripped" from and "burnt" onto a compact disk, it can be saved on any device that has a hard drive, it can be easily transported over the Internet, and it can be swapped between people who have never met. And it is invisible. It is ephemeral not corporeal, it can but need not be attached to a physical medium such as a CD, it cannot be seen or touched but it can be heard. Being binary in its construction it never degrades, and it has eroded the difference between original and copy, making the term high fidelity less meaningful. Though mobile externally, internally it is absolutely static and unchanging. Bruno Latour's concept of the "immutable mobile" (1986) as something that can effortlessly traverse space without being transformed in the process works as an appropriate metaphor (with the appropriate software, though, MP3s can be modified). This inherent immutability of the MP3 file that allows it to be reproduced numerous times without loss of sound quality has also allowed music to navigate easily across various locations (computers, MP3 players, compact disks, the Internet).

The MP3 player

Considering the popularity and success of portable music players like the Walkman and Discman, the debut of a mobile MP3 player was imminent. The first portable MP3 player to be released in the United States was the Rio, from Diamond Multimedia, in 1998. Since then, many more have appeared on the market, and the iPod was first announced at a news conference in Cupertino, California, on October 23, 2001. In the short time since its introduction, the iPod has become one of Apple's most popular products, with over 750,000 units sold in its first fiscal quarter of 2004. A crucial product in the emerging landscape of digital music, the iPod has become fetishized rather quickly. Described as a cult object in the *New York Times*, sold at the rate of two every minute, winner of several design awards, this cool new product reflects the styles, attitudes, and new patterns of behavior of several user groups. Through this process, it has become a cultural commodity that has changed how music is shared, transported, distributed, stored, and consumed.

Indexing personal stereo use

Portable music players have been the subject of numerous studies, resulting in many articles and two books, one titled *Doing Cultural Studies: The Story of the Sony Walkman*, by Paul du Gay, Stuart Hall, Linda Janes, Hugh Mackay, and Keith Negus (1997); and the other, *Sounding Out the City: Personal Stereos and the Management of Everyday Life*, by Michael Bull (2000). The study performed by du Gay *et al.* applies a model they refer to as the circuit of culture that includes five processes: production, consumption, regulation, representation, and identity. These five, which roughly form the chapters of the book, can be employed in performing similar analyses on other cultural artifacts as well. Michael Bull, on the other hand, investigates the nature and role of the mobile auditory experience, technology, and personal stereos in the management of everyday life. Bull offers a typology as a means of cataloging the reasons why personal stereos are used, some of which are direct responses to the environment. These gadgets offer a means by which undesirable and loud city sounds can be replaced by a personal soundscape. They also allow people to withdraw into themselves from the discomfort, hostility, and insecurity caused by the omnipresent urban multitude via the activity of privatized listening. Sometimes personal stereos (and especially, in this case, the music played on them) provide an esthetic experience that enhances the outside environment. On the other hand, Bull's studies have shown that people may have little interest in the environment, and therefore rely on their Walkmans to make the journey more engaging through music-inspired sensations and memories. In an environment that can be annoying or meddlesome, the headphones act as shields from unwanted human contact. Many other nuanced uses for the Walkman are discussed by Bull, most of which can be understood as strategies of dealing with the self and the environment. The multiple uses of MP3 players such as the iPod can be explained using Bull's typology, but they offer added advantages not previously experienced by fans of portable music.

Either too bulky or awkwardly shaped, the Walkman and the Discman do not fit easily into pockets, and they have to be carried around in the hand or in a bag. They also require media (cassette tapes and CDs) which hold the music. MP3 players are smaller and do not need external media, as they store music files on their internal hard drives. This has added tremendously to their transportability and overall convenience, an aspect that is promoted through iPod advertising. The Apple Computer website regularly features commercials in the form of Quicktime videos of the products and/or services, one of which presents Hamilton Morris, a student who situates the iPod within his daily activities.

> It's kind of a pain to carry around a whole lot of CDs, so I try and stick all of them inside one case, but that never works cause then it puts too much pressure on the little door and [it] always gets cracked and split[s] open. You have these heavy textbooks that are banging against them all day and CDs spill out and make it scratch by all the crumbs in your bag . . . I love my iPod. A hundred CDs in something that's, you know . . . that big [gestures and laughs]. So much music . . . it makes me feel powerful. Hamilton Morris, I'm a student.
> (http://www.apple.com)

Through Hamilton, Apple has created the identity of one of the typical users of an MP3 player: a student always on the move with plenty to carry around, interested in keeping a lot of music at hand, not very organized, etc. The identity of the product itself is shaped through the monologue which offers a comparative analysis of old media and new, referring to the hassles of one and the convenience of the other, the clumsy nuisance of one and the easy transportability of the other, the fragility of one and the power of the other. This creates a differentiation for the product within its milieu. Interestingly, the iPod is never shown in this clip; the emphasis is entirely on the experience it provides in the context of everyday life as demonstrated in references to CDs, the CD case, textbooks, and, of course, the crumbs in the bag.

The iPod as environment

The transportable environment created by the iPod may be understood as a site where body, technology, and space come together to construct an adjusted meaning of the private. As you walk the city, iPod in hand and signature white ear buds plugged in, this private bubble travels with you like a shadow, ever present and close by. In his essay "To Each Their Own Bubble," Michael Bull refers to the personal space created by personal stereo users as "non-spatialized conceptual space," and the process of generation of this space as a form of "colonization" (2004, p. 284). He further notes that this space is more ephemeral than geographical by necessity, largely because of the difficulty of substantiating and negotiating personal space in today's urban environments.

Some of the concepts developed by philosophers Gilles Deleuze and Felix Guattari may also be successfully applied as models in understanding iPod environments. For example, the environment created by the iPod may be seen as a Deleuzian assemblage. In conversation with Claire Parnet, Deleuze describes assemblages as entities that have four components or dimensions. Assemblages are "states of things" that are suitable to the individual, they are "little statements" that represent style, they "implicate territory," and also "implicate deterritorialization" (Deleuze, 1977). Though these discussions of assemblages between Deleuze and Parnet occur within the context of the concept of desire, the four components may be employed as metaphors in understanding the territory set up by the iPod environment. A transportable iPod environment is a state of being donned by the user in creating a suitable protective atmosphere. It also makes a statement to the public realm about the privatized state of mind of the iPodder. Environments, whether real or virtual, claim territory by their very existence. In this case, a boundary is set up when the headphones are slipped on, and is maintained as long as they stay on. However, this is a fragile, permeable boundary that can be shattered or breached by external forces and interruptions. As the margins of this territory are porous and because the territory is constantly on the move, deterritorialization is easily achieved. The iPod environment, as an assemblage, also contains music. The music that fills it provides a form of utopia. "By circling people, by enveloping them – as inherent in the acoustical phenomenon – and turning them as listeners into participants," music adds dimension to the assemblage (Adorno, 1976, p. 46). Therefore, one may describe the transportable environment of the iPod as an assemblage with multiple dimensions, real and virtual, enveloping but permeable, territorialized but moving, and entirely soaked in sound.

Mobile privatization

If the consumption of music is understood as a leisure activity connected to the private domain of the domestic, the iPod mobilizes it and takes it out into the street. To have music close at hand at any given time in any given space in a day is a crucial need for many, as is testified by the proliferation of iPod accessories on the market, such as waterproof carrying cases, arm bands, the iTrip (which is a transmission device that allows the iPod to be heard over a radio station), chargers for the car, small speakers for the office, etc. Recently, Apple along with BMW has introduced a special integrated adaptor for the iPod that allows it to be connected to the existing audio systems in some BMW vehicles. Through these add-on devices, the gadget enlarges its environment and extends its reach into spaces previously inaccessible. In his discussion of technology and society, Raymond Williams talks about the "two apparently paradoxical yet deeply connected tendencies of modern urban industrial living: on the one hand, mobility, on the other hand the more apparently self-sufficient family home." He refers to this simultaneously "mobile and home-centred way of living" as a form of mobile privatization, and to him these represent "contradictory pressures of this phase of industrial capitalist society," that need some form of resolution (Williams, 1975, p. 26). In the 1920s, people who lived in these self-sufficient family homes needed improved communications with the outside, and that was provided through private transportation and new consumer technologies such as radio and television. However, one of the limitations of these media was their centralized transmission and individual reception. Though television and radio offer several channels and stations for specific audiences using the concept of narrowcasting rather than broadcasting, they are not customizable. Media corporations rather than people control the programming, and therefore the environments they create cannot be tailored to the specific needs of individuals. Not based on the model of transmission for public use, MP3 technology has solved this problem to a certain extent by giving users control of the type and amount of music that they may load on the players, its organization into personalized playlists, its categorization, etc. iPods function as a means of managing one's place in the location where one might find oneself. Often used while waiting or during travel between environments, they also serve to rearrange users' experience of time.

The iPod affords the possibility of creating special "soundtracks" to accompany routine activities such as riding or driving. The studied creation of playlists that allow one-to-one correspondences between music and activity in terms of genre (for mood generation) as well as length of time (to match the duration of a run, for example) is a signifying practice that makes sense of the technology and the object, as well as the activity. The following excerpt from an interview I conducted with an iPod user makes it clear that it is remarkably easy to create special lists of songs with varied time durations for specific activities.

> K: This [the iPod] is something that pretty much accompanies me anytime that I am not engaged with other people, when I am not, you know, sitting and talking, or, you know, in a place where, where I have to pay attention to whatever, you know the traffic or . . . but actually most of the time I've got it on my bike or you know when I am back and forth between here and there. And the way that I've got the playlists arranged, it's, you know, like, six or seven songs, segments which is basically the distance that takes me to get from here to there. So I've got these, you know, these short little tracks, that I can listen to while I am riding, or whatever and it's all set up like soundtrack wise, you know . . .
> Interviewer: Hmm, so do you have special playlists depending upon . . .?
> K: . . . what I'm doing? Yeah . . . absolutely. I've got soundtracks to everything I do. I think everybody does, though, you know.
> Interviewer: Can you talk more about that?
> K: Sure. Absolutely.
> Interviewer: Or you can give me an example, or something . . .?
> K: Sure, I'll make examples. There's certain music I'll listen to when I'm doing certain activities, and I think that's probably pretty common . . . I've got, you know, collections on there which are strictly hip-hop, that you know, I'll listen to in certain moods, certain activities . . .

> and I've got, you know, old like classic rock stuff, that I'll listen to when I'm doing other things, you know when I'm like at the gym or whatever, you know, it's just certain things that sort of fit the bill . . .
> (Interview 2)

Such technological individuation of musical experience in relation to everyday activities firmly locates objects into people's lives. The processes of searching, finding, and downloading individual music files are entirely private and decentralized, and require no social intercourse. Though not all music acquisition occurs through this channel and record stores are still in business, it is significant to note that the only communally visible activity that an iPodder may engage in might be that of listening. This reveals a remarkable reversal in private and public activities.

iPod communities

Virtual environments such as electronic forums and weblogs may be understood as locations where several private environments can interact. As contact in the public domain erodes, it grows with incredible speed in a virtual one, adding a second-order envelope to the existing transportable environment. The boundaries between the private and public are therefore permeable and elastic; they stretch and they overlap, defying definition and categorization in the creation of hybrid spaces. iPodLounge is such a space. It promotes the individual narrative within the context of a mutually agreeable social structure. www.iPodLounge.com was born within a month after the iPod was released in 2001. Conceived by Dennis Lloyd, Dennis Martin, and Jason Meade, this is a website "dedicated to the iPod enthusiast." iPodLounge is not a commercial site sponsored or supported by Apple Computer, but a communal space with news, reviews, forums for discussions, technical advice, etc. "for all things iPod." This website may be seen as an alternative and more developed form of the iPod's transportable environment. It therefore becomes a meta-environment, which readjusts meanings of private and public. As a meta-environment in the public cyber domain, it is accessible to all (but known only to the initiated), and through a forum called "Your iPod Stories," it encourages exchange of private narratives of use.

The privatization of most activities connected to music consumption might explain the mutual identification experienced by some iPod users in public spaces (called Pod spotting) and the appearance of Internet communities such as iPodLounge and iPodding might be interpreted as a need to reconnect with society. The transportable environment, then, at least partially, becomes a means to publicize rather than privatize, an instrument to engage rather than disengage.

Individual and society

As an object designed primarily for personal, mobile use, the iPod can be critiqued in terms of the relationship between the individual and society. Raymond Williams limits this liaison to a simple model of the individual's conformity or nonconformity, and of the society's attitude to either of these courses. We have a number of names for conformity, which enable us to approve it as "responsible" and "law-abiding," or to condemn it as "timidly conventional" or "servile." We also have a number of names for nonconformity and some of these, such as "independence" and "the free spirit," are approving, while others, such as "lawlessness" and "eccentricity," are damning (Higgins, 2001).

Users of mobile stereos and other gadgets such as mobile phones that express private behaviors in public space are often labeled as self-engrossed, rude, and lawless. As du Gay noted, on the London Underground Walkman users have been given contemptuous looks as if to say "SCUMBAG!

LOW-LIFE! LOSER!" (capitals in the original text), they have been fined, and treated as a menace to society (du Gay *et al*., 1997, p. 144). The individuals in these cases have been perceived as disruptive and destructive to the social order. On the other hand, iPod users have been offended by people who interrupt their listening. Using it as a device to insulate themselves from preying salespeople, iPodders saunter through shopping malls cocooned in the safety of their auditory environments, and are not pleased to have to stop their players. The following exchanges at the iPodLounge website demonstrate how this environment may be disturbed.

> People who have always used Walkmen might already be keen to this, but I've realized that annoying salespeople are much more reluctant to bother you if you have your earbuds in your ears. Over the holidays my iPod enabled me to have a relatively hassle-free and private shopping experience wherever I went. I guess people who aren't as annoyed by non-stop "Can I help you find something?" as I am might not find this tip helpful, but for those who leave a store once a salesperson pounces (annoyed with yourself for not having shopped online), give it a try.
>
> Once at Brookstone a lady asked me if I needed help, and I thought she was talking to my grandma who was behind me. I didn't look or answer and then she yelled at me, "Excuse me sir. Do you want any help?" It was all very funny.
>
> Wow this has never happened to me. I would be annoyed if they told me to take them off. Interrupting good music.
> (http://www.iPodLounge.com/)

The individual in this case may be perceived as being menaced (or amused) by society. However, the practice of individual listening in social circumstances has become so prevalent that it may be easily classified as public behavior, making it problematic to clearly and distinctly separate the individual from society.

Permeable boundaries (public–private)

Many aspects of mobile behavior are heavily technologized, and this has made it easier to create private domains to withdraw into while being on display in the public realm. The ubiquity of mobile technology can be witnessed in any urban streetscape, which is full of people walking while talking on cell phones, jogging while listening to music on their MP3 players, or sitting at cafés while working on their laptops. Since the Industrial Revolution, many scholars and critics have studied the effects of capitalism, modernity, and technology on urban environments and individuals. Georg Simmel emphasizes the struggle between the private and the public in his description of the urban experience; it is a world of strangers where contact is awkward and boundaries are required.

> Since contemporary urban culture, with its commercial, professional and social intercourse, forces us to be physically close to an enormous number of people, sensitive and nervous people would sink into despair if the objectification of social relationships did not bring with it an inner boundary and reserve.
> (Simmel, 1997)

Such boundaries signify private environments that ease urban angst, and the presence of gadgets, which can occupy time, space, and attention, solidify individual positions within these spheres. Richard Sennett (1974) has observed the erosion of public life, and the need for privatization of experience from a chaotic urban environment.

Though the act of private listening in the public domain may be a means of dealing with the discomfort and alienation experienced by individuals in modern society, there is reason to believe that the chasm between these realms is not quite as deep everywhere. In the case of iPod use, there is a marked sense of communal association and public connection on occasions when one encounters another iPod user.

> Here in NY I see folks with pods daily. Usually the white earbuds are a dead giveaway, and if I have mine in as well, the fellow 'podder and I usually lock eyes for a moment (which in NY is a lot!).
>
> I saw a dude on the bus here in Philly yesterday with the white buds in his ears . . . I just kinda gave him "the nod" and a smile. (http://www.iPodLounge.com/)

Though this link establishes contact with another individual rather than the public, it does unwrap the user's environment for a brief moment. As is demonstrated in this exchange on the iPod website, a shared privacy is sensed in such situations of mutual identification, where two environments accidentally meet.

Conclusion

Digital music cultures have created new products, new practices, and new means of dealing with urban spaces. The iPod, one of Apple Computer's most popular products today, has inserted itself into people's lives and has given them new strategies for dealing with the public domain. The iPod's environment is an assemblage that allows and supports various behaviors. It mediates between the individual and the society, it negotiates the meaning of the private in public spaces, and it has led to the evolution of virtual communities. As an object that embodies the concept of mobile privatization, it represents the present and future of transportable music consumption.

References

Adorno T. (1976) *Introduction to the Sociology of Music*. New York: Continuum Press.

Adorno, T. (1991) *The Culture Industry: Selected Essays on Mass Culture*. London: Routledge.

Alderman, J. (2001) *Sonic Boom: Napster, MP3, and the New Pioneers of Music*. Cambridge: Perseus.

Appadurai, A. (1986) "Introduction: Commodities and the Politics of Value." In A. Appadurai (ed.), *The Social Life of Things: Commodities in Social Perspective* (pp. 3–63). Cambridge: Cambridge University Press.

Arthur, C. (2003) "Mr. iMac Wins Design Prize for Banishing Beige Boxes,' *Independent*, June 3, p. 7.

Breen, C. (2002) *Secrets of the iPod*. Berkeley: Peachpit Press.

Brow, J. and Duguid, P. (1993) "Rethinking the Border in Design: An Exploration of Central and Peripheral Relations in Practice." In S. Yelavich (ed.), *The Edge of the Millennium: An International Critique of Architecture, Urban Planning, Product and Communication Design* (pp. 174–189). New York: Whitney Library of Design.

Bull, M. (2000) *Sounding Out the City: Personal Stereos and the Management of Everyday Life*. Oxford: Berg.

Bull, M. (2004) "'To Each Their Own Bubble': Mobile Spaces of Sound in the City." in A. Couldry and A. McCarthy (eds), *Mediaspace: Place, Scale and Culture in a Media Age* (pp. 275–293). London: Routledge.

Chambers, I. (1982) "Some Critical Tracks," *Popular Music,* 2: pp. 19–36.

Cooper, G. (2002) "The Mutable Mobile: Social Theory in the Wireless World." In B. Brown, N. Green, and R. Harper (eds), *Wireless World*. London: Springer-Verlag.

Deleuze, G. (1977) *Dialogues/Gilles Deleuze, Claire Parnet*. Paris: Flammarion.

DeNora, T. (2000) *Music in Everyday Life*. Cambridge: Cambridge University Press.

du Gay, P., Hall, S., Janes, L., Mackay, H., and Negus, K. (1997) *Doing Cultural Studies: The Story of the Sony Walkman*. London: Sage Publications.

Haug, W. (1986) *Critique of Commodity Aesthetics: Appearance, Sexuality and Advertising in Capitalist Society*. Cambridge: Polity Press.

Heskett, J. (2002) *Toothpicks and Logos: Design in Everyday Life*. Oxford: Oxford University Press.

Higgins, J. (2001) *The Raymond Williams Reader*. Oxford: Blackwell.
Highmore, B. (2002) *Everyday Life and Cultural Theory: An Introduction*. London: Routledge.
http://www.apple.com
http://www.iPodLounge.com
Jones, S. (2000) "Music and the Internet." *Popular Music,* 19/2: pp. 217–230.
Kunkel, P. (1999) *AppleDesign: The Work of the Apple Industrial Design Group*. New York: Graphis.
Latour, B. (1986) "Visualization and Cognition: Thinking with Eyes and Hands." *Knowledge and Society*, 6: pp. 1–40.
Lefebvre, H. (1999) *Everyday Life in the Modern World*. New Brunswick: Transaction.
Mitchell, W. (2002) "E- Bodies, E- Building, E- Cities." In N. Leach (ed.), *Designing for a Digital World* (pp. 50–56). Chichester: Wiley-Academy.
Sennett, R. (1974) *The Fall of Public Man*. New York: Norton.
Simmel, G. (1997) "The Metropolis and Mental Life" in D. Frisby and M. Featherstone (eds), *Simmel on Culture: Selected Writings*. (pp. 174–186). London: Thousand Oaks.
Williams, R. (1975) *Television: Technology and Cultural Form*. New York: Schocken Books.

Oil and Water: Offshore Architecture

Justin Beal

The program of the offshore oil rig as it has been defined over the last fifty years has demanded a new system of design solutions that combine nautical, architectural, and technological models into a new hybrid form. Today, there are thousands of active rigs drilling for oil in the North Sea, the Niger Delta, the Persian Gulf, the Campos Basin, and other locations worldwide. This research (which has been undertaken as an accompaniment to the art project *Everything is Going According to Plan: Parts I–III*, a series of digital composites comprised of hundreds of images of offshore drilling installations) examines the offshore oil installation as a unique cultural and architectural typology through a general objective analysis as well as a catalog of several specific rigs. It defines the parameters of this typology as it has developed under the influence of the functional demands of offshore drilling, the psychological pressure of such a dangerous undertaking, the mythology of the rig in popular culture, and the privileged role of oil companies in the global market. Ultimately, the rig as it relates to past and present models of futurist design is considered, leaving unanswered, however, the question of how this typology might inform the design of tomorrow.

The mythology of offshore oil

> The great leviathan is that one creature in the world which must remain unpainted to the last. True, one portrait may hit the mark much nearer than another, but none can hit it with any very considerable degree of exactness. So there is no earthly way of finding out precisely what the whale really looks like. And the only mode in which you can observe even a tolerable idea of his living contour, is by going a whaling yourself; but by so doing, you run no small risk of being eternally stove and sunk by him.
> (Melville, [1851], 1971, Ch. 55)

> Rare are those who've seen the platforms, and rarer still those who've recounted what they've seen.
> (Béguin, 2001, p. 122)

Just as the experience of early maritime life became generally known through rough etchings and vivid literary descriptions, the phenomenon of the offshore drilling installation has become universally recognizable not through the public's first-hand experience of such rigs but through the medium of film. Everyone has seen oil rigs on film. The mythic image is a towering high-tech super-structure, painted with vibrant colors and decorated with helipads, derricks and cranes, set against the sublime background of the open sea. Stacked platforms teem with workers bearing titles like roughneck, toolpusher, and roustabout as waves crash against concrete pylons at the base of the platform and flames spew from steel-truss flare-booms at the top. The Hollywood blockbuster *Armageddon* (1998) introduces its protagonist Harry Stamper (Bruce Willis) with beautiful sweeping helicopter shots of a rig in the South China Sea. This image, coupled with relentless references to *The Right Stuff* (1983), frame the rig as a proto-space station (a specific point to which I will return in the conclusion of this text). *Armageddon* perpetuates the

4.1 *Everything is Going According to Plan: Part II* (detail)

hyper-macho myth of the roughnecks who, with their unique combination of endurance, ingenuity, and huge drills, are recruited to save the world. The rig sets the stage for a similarly exaggerated portrait of masculinity when the former CIA operative/ extreme firefighter Forrest Taft (Steven Seagal) attempts to save the Alaskan wilderness on behalf of its native inhabitants in *On Deadly Ground*. In addition to providing irresistible action-sequence mise-en-scène, the rig in *On Deadly Ground* is given a political identity as the menacing fortress of the evil oil executives. More chilling, if realistic, representations of the rig include Lars Von Trier's *Breaking the Waves* (1996), which depicts the extreme working conditions on a North Sea oil rig with documentary-like realism, and Michelangelo Antonioni's *Red Desert* (1964), in which the rig appears as a more abstract symbol of modern society's post-industrial anxiety and isolation. In all of these films, the myth of the oil rig is perpetuated through an emphasis on the extreme environments and sublime structures of offshore installations. Yet the reality is often more exotic than the myth that has come to surround it. In the seemingly absurd plot of the James Bond film *Diamonds Are Forever* (1971), a rig off the coast of Baja, California, is the secret control station of the evil mastermind Ernst Blofeld's diamond-gilded satellite. Twenty-five years later, in Long Beach, California, Boeing and the Norwegian company Knaerner announced plans for a $500-million project to convert a 31,000-ton decommissioned oil rig into an offshore satellite launch platform ("The Lay of the Land," 1997).

The artificial island

> Most sea stories are allegories of authority. In this sense alone politics is never far away. The ship is one of the last unequivocal bastions of absolutism, regardless of the political system behind the flag that flies from the stern, or behind the corporate logo behind the flag that flies from the stern. This makes ships all the more curious and anachronistic in an age proclaimed to be one of the worldwide democratization. Ships function both as prisons and as engines of flight and escape.
> (Sekula, 2002, p. 118)

> If we think, after all, that the boat is a floating piece of space, a place without a place, that exists by itself, that is closed in on itself and at the same time is given over to the infinity of the sea . . . the ship has not only been for our civilization, from the sixteenth century until the present, the great instrument of economic development . . . but has simultaneously the greatest reserve of the imagination. The ship is the heterotopia par excellence.
> (Foucault, 1986, p. 27)

In terms of language, the rig can only be understood in nautical terms – booms and ballasts, heads and holds, decks and docks, keeps and cabins. Indeed, by design, the rig is essentially a mobile structure. The extent of this mobility ranges from short-term drilling boats tethered to the seabed with steel umbilical cables to nearly permanent single-use platforms on massive concrete pylons. Yet, because of the limitations of offshore deepwater construction, every oil installation necessarily begins as a vessel. A massive armature is towed into place and assembled gradually on site. In some cases a rig will travel extensively before coming to rest at its final installation. Like cargo ships, which are regularly overhauled, renovated, and rebuilt in various ports, each rig wears the marks of its various travels. The *Petrobras-36* – the largest platform in the world when it sank in 2001 – was originally the *Spirit of Columbus*, built in Japan and modified in Canada before settling in Brazil's Campos Basin. It took six weeks to transport another Brazilian rig, the 40,000–ton *Petrobras-40*, from Singapore via the Cape of Good Hope to the Guanabara Bay in 2000 (Dockwise News Desk, 1999). It is this simultaneous potential for mobility and practical immobility that

4.2 Shell *Auk*: a platform in the North Sea

4.3 Pride International's *Pride Brazil* semi-submersible Platform

4.4 Shell/Esso *North Cormorant* platform in the North Sea

begins to define the complex identity of the rig as a site – as both a vehicle and a structure. On one level the rig occupies the small plot of ocean floor, which it punctures while its de facto site is the smooth space of the water above, which it just barely surfaces. Unlike a vessel, which can move with or against the sea, the rig remains stationary while the surface it occupies constantly fluctuates. The rig is a static figure on a dynamic ground. This idea alone requires a radical rethinking of the traditional architectural notion of site. The rig is an island unto itself – neither building nor boat.

The potential for mobility combined with a constantly changing workforce gives each rig a unique political autonomy. Workers are typically helicoptered in for three consecutive weeks of twelve-hour shifts followed by three weeks on shore. These workers generally come from remote locations. It would not be uncommon, for example, for a Texas-based conglomerate to employ British workers on a Norwegian rig in the Niger Delta. Most rigs, regardless of their position, operate not under a flag but a logo, and the ultimate allegiance of the crew is corporate not patriotic. This apparent autonomy raises questions of legal jurisdiction: what body presides over offshore platforms? And what power do they have to exercise that authority? Even a rig stationed in national waters will be more closely monitored by its parent corporation than by any local authority. In her essay, "Legal Issues of Ocean Cities," Cordula Fitzpatrick offers the following definition of artificial islands:

> Artificial islands and installations are man-made, surrounded by water from all sides, above water at high tide, supposed to stay at a specific geographical location for a certain span of time, and stationary in their normal mode of operation at sea . . . the exclusive right is only granted if the installation serves an economic purpose . . . the State may provide for so-called safety zones for artificial islands and installations located outside of territorial waters, as artificial islands and installations are not entitled to a territorial sea or other maritime zones. (Fitzpatrick, 1999, p. 3)

Despite being engineered to withstand 80-foot waves and icebergs twice its size, the rig offers little hope of swift movement and its site is one of extreme vulnerability. The rig is a space fraught with all the claustrophobia and anxiety associated with maritime life, only compounded by the fact that, being essentially stationary, its crew does not have the luxury of escape. Like a ship moored at permanent anchor, it is locked in place and between ports, unable to flee from inclement weather, but also from unwelcome visitors. The identity of the artificial island of the rig as the problematized mascot of the oil industry makes it a target of environmental and economic activists and a highly politicized site of piracy and protest. So, even if rigs are subject to international law and corporate vigilance, they remain in a position of astonishing vulnerability.

This is particularly evident in the troubled history of rigs drilling off the West African coast. The latest in a series of hostage situations on rigs in Nigeria occurred in April 2003: striking oil workers held ninety-seven people hostage on several rigs simultaneously (CBSNews.com, 2003). In May 1998, 120 to 200 young protestors took over Chevron's *Parabe* oil platform about 15 kilometers off the Niger Delta. The unarmed youths staged a protest over environmental damage caused by oil spills, the disruption of communities' water supplies, the ruining of local fishing and farming trades and the failure to employ locals. Two protestors were killed and thirty injured when government troops intervened (AFROL News, 2002). In August of 2001, forty young protestors from Ilaje communities in Ondo State besieged the *Trident-8* oil rig off the coast of Nigeria and held ninety oil workers captive for nearly four days. By barricading the helicopter landing pad, the protestors were able to block the oil company's only means of access to the rig. The protesters were demanding contract jobs from *Trident-8*'s owner, Houston-based Transocean Sedco Forex. The situation was resolved when community elders were called in by Shell to ask the youths to leave after three unsuccessful days of negotiation. The protestors' demands were never met. However, what makes the story of the *Trident-8* particularly poignant is the fact that the news of the incident was only released after the situation was resolved. According to BBC correspondent Dan Isaacs, "incidents such as this are not uncommon in the country's troubled oil-producing region, but oil companies rarely provide details of kidnap situations such as this until they have been resolved, for fear of jeopardizing negotiations" (Isaacs, 2001). The continuing situation in Nigeria suggests that the corporations maintain ultimate control. The power oil companies exert through control over communications and access makes it hard for any other authority to claim sovereignty, regardless of what the law states.

The disaster

> The building is effaced in favor of the inhabited machine. Having become more of a hindrance than a help, the envelope disappears. Bits of machinery and living quarters exist side by side within a restricted perimeter and tend to intermingle. In as far as the interdependence of built objects – and that with their terrestrial container – functions through the intermediary of this envelope, the platform manifests a clean break with this great and ancient family of buildings.
> (Béguin, 2001, p. 123)

> The philosophy of the majors in recent years has been to run the rigs for cash, and repair things as they've failed, rather than perform routine maintenance . . . in the first few years after that philosophical change, you get a profit boost, as not everything fails immediately. But after a while that catches up with you.
> (David Hobbs quoted in Timmons, 2003, sect. W, p. 1)

The story of offshore oil exploration, at its foundation, is one of high-stakes technological and economic venture. Rigs are, first and foremost, machines. The design of everything on a rig is dictated by economic constraints, wherein the primary objective is to streamline operations and maximize output. Rigs are engineered, not "architected," and each platform is in fact a very specialized machine – essentially a factory equipped to locate, extract, store, and, in some instances, process oil. It is stripped of anything non-essential – walls, roofs, foundations – giving it the appearance of a refinery or factory that has been turned inside out. With the exception of the soundproofed living quarters and iconic octagonal heliports, which often appear to be clipped to structures as an afterthought, everything on the rig has a specific mechanical function.

From a historical perspective, maritime exploration and oil speculation have always been fueled by disproportionate economic interest and high-risk investments. Today, images of offshore rigs decorate currency in Nigeria, Saudi Arabia, Azerbaijan, Malaysia, and Trinidad and Tobago. The promise of extraordinary returns has justified otherwise unthinkable undertakings and unprecedented risk to human safety. This idea, perhaps more than any other, perpetuates the myth of the oil rig as a site of extreme danger. This is only exaggerated by the extreme size of these installations. When it was anchored in 1982, the 900,000-ton Norwegian rig, *Statfjord-B*, was the heaviest object ever moved (Béguin, 2001, pp. 122–123). After an article on the *Statfjord-B* was published in *Scientific American*, some people speculated that if the sortie of tugs had lost control of the rig as it left the protected fjord, the impact of a single pylon hitting the ocean floor could have caused a minor earthquake on shore. Though

4.5 *Petrobras-36* platform listing severely before sinking in March, 2001

4.6 Rig sinking in the Khafji oil field in the Persian Gulf

4.7 Currency from Saudi Arabia

4.8 Currency from Nigeria

the *Statfjord-B* arrived on site unscathed, the speculation suggests the potential magnitude of accidents on such large-scale structures.

The Brazilian rig *Petrobras-36* – once the largest drilling platform in the world – was 250 feet long, 250 feet wide, 150 feet tall and weighed over 30,000 tons. With a potential daily oil production of 180,000 barrels, it was capable of producing approximately a tenth of Brazil's national oil output. On Tuesday March 20, 2001 the *Petrobras-36* sank in the ultra-deep waters of the Campos Basin in the Roncador oil fields about 80 miles off the north-eastern coast of the state of Rio de Janeiro. Eleven of the rig's fire-fighting crew died in the accident. According to the Petrobras public relations department, the tubes that connect the wells to the platform and the pipelines were cut using robots that operate at the bottom of the sea (narrowly avoiding a major environmental catastrophe). Though the sinking of the *Petrobras-36* was the most publicized Brazilian offshore accident in decades (owing in part to spectacular aerial photography), a total of ninety-three Petrobras workers have died in smaller incidents on the job in the last four years (Forbes.com, 2001). Despite these statistics, from an economic perspective, the returns seem to justify the risk. Petrobras is Brazil's largest company with annual revenue in excess of $16 billion. Before it sank in 2001, the *Petrobras-36* rig had the capacity to produce upwards of $6 million worth of oil a day. That means that if the platform were a small country it would have an annual income of 22 million dollars per person or 1.2 million dollars per square foot (Forbes.com, 2001). In the simplest terms, the economics are always pushing the technology beyond its logical limits in the interest of increasing production.

It is the threat of disaster that is the overlap between the ship and the machine. Neither the nautical culture nor the architecture of oil can be separated from the influence of the imminent disaster. The technology of oil extraction offers the threat of explosion while the ship's hull is in constant danger of rupture. The volatility of high-pressure pipelines threatens to blow the rig out of the water while the expanse of the ocean threatens to pull it back in. Both the isolation and the double threat of fire and water create a space of heightened anxiety. The environmental and economic repercussions of even a minor accident can be devastating. The rig is constantly influenced by the possibility of imminent catastrophe – explosion or submersion. In August 1984, thirty-six workers drowned and seventeen were injured in an explosion and fire on a Petrobras oil-drilling platform in the Campos Basin. In March 1992, eleven men died when a French-built Super Puma helicopter, carrying fifteen workers and two crewmen, plunged into the North Sea seconds after taking off from the *Cormorant Alpha* platform (CNN.com/World, 2001). The worst catastrophes on record include the capsizing of the *Alexander Keilland* oil rig in the Ekofisk field of the North Sea where 123 people died in 1980 and the Valentine's Day 1982 sinking of the exploratory oil rig *Ocean Ranger* in the North Atlantic, which killed eighty-four people (Saunders, 2002). Today, the average age of offshore workers in the North Sea is 50 (Timmons, 2003). The most devastating and best-documented accident on record was the 1988 explosion of Occidental Petroleum's *Piper Alpha* oil rig near Aberdeen in the North Sea. One hundred and sixty workers died (of 229 on board). Despite the fact that crew members had been complaining about a heavy odor of natural gas on the rig for several days before the accident, the company was not charged with any wrongdoing. The accident is often attributed to cost-cutting which followed the slump in oil prices in 1986 (BBC News Online Network, 1999). In an interview with the *New York Times,* Jake Molloy of the offshore workers' union recalled:

> Everyone that was working offshore back then knew it was only a matter of time before we had a big incident, and many pointed to Piper as the place most likely to suffer ... now

they're saying, "it's only a matter of time before there's another accident in the North Sea, and it could be anywhere."
(Timmons, 2003, sect. W, p. 1).

The story of offshore exploration as it is presented by media outlets gives a series of negative impressions, inextricably attached to the possibility of catastrophe. The only time rigs appear in the news is when something goes terribly wrong, and even then coverage is generally reserved to the financial outlets. This portrait of the rig presented through the news media is surprisingly consistent with the myth of the rig as it appears in film – in more than half of the films mentioned in the introduction the rig ends up in flames.

The megastructure

It could be that one day in the future the silhouette of the offshore platform typifies an entire epoch – in economic, technical, morphological terms – just as yesterday and the day before were typified by the windmill and the factory.
(Béguin, 2001, p. 123)

Continental civilization has constantly spread bloody strife among that mankind fated to live on land. It may not be too much to say that continental civilization has been no more than a history of conflict. And today the world is being daily threatened with the final confrontation between the two continents . . . the marine city will be born as, and must be made into, a city truly contributing to human society.
(Kikutake, 1959, in Eaton, 2001, p. 220)

The myth of offshore exploration generally casts the rig as the embodiment of corporate greed and environmental irresponsibility. Rigs are associated with danger, with pollution, with globalization, yet there is an amazing potential within the form to be totally autonomous and to adapt and expand with remarkable ease. If it is possible to detach the identity of the rig from its environmental and economic role, it can be viewed instead as a unique realization of some of the most visionary architectural ideas of the last fifty years. At its essence, the rig bears a curious relationship to the hypothetical environments proposed by the so-called megastructuralist drawings of groups like the International Situationists, the Metabolists, and the Italian Futurists as well as the evolving extraterrestrial architectures of space stations and satellites.

4.9 Shell/Esso *Fulmar*: A platform in the North Sea

Offshore oil exploration was only a nascent industry in the 1960s when megastructuralist images began to emerge in European architectural circles. Works like Yona Friedman's *Spatial City* series and Archigram's *Plug-in City* proposed a utopian vision of a truly three-dimensional and infinitely expandable urban environment, bearing an uncanny likeness to the technology being explored by the North Sea oil companies at the same moment. Ralph Wilcoxon defined the megastructure in 1968 as:

Not only a structure of great size, but . . . also a structure which is frequently: 1 constructed of modular units; 2 capable of great or even 'unlimited' extension; 3 a structural framework into which smaller structural units (for example, rooms, houses, or small buildings of other sorts) can be built – or even 'plugged-in' or 'clipped-on' after having been prefabricated elsewhere; 4 a structural framework expected to have a useful life much longer that that of the smaller units which it might support.
(Wilcoxon, 1968, p. 2)

The emerging utopian vision was a metropolis as a machine; an autonomous, adaptable urban environment coupled with the hope that, in the future, technology would advance to such a degree that a minimum amount of human labor would be required to maintain a massive city. At the same time as the offshore oil industry in the North Sea was slowly evolving, megastructuralist drawing championed the ideas of adaptability and flexibility, plugging in and clipping on. The most developed of these projects was the extensive catalog of drawings, photographs, and models that comprised International Situationist artist Constant Nieuwenhuis' *New Babylon* project. The technological and visual similarities between the elaborate structures of the *New Babylon* and the clustered groupings of oil rigs common today are striking – the raised platforms, the vibrant colors, and flexible space-frame systems. The visual language of contemporary rigs – composed of an infinitely expandable pastiche of parts rather than an architecturally organized whole – can be described as a sort of structural manifestation of the Situationist notion of *détournement*. Moreover, the rig as an autonomous floating city seems to exceed even the most far-fetched futurist ideas of utopian urbanism (the non-permanence of the rig gives a new meaning to the other Situationist tenet, *dérive*, by setting the entire environment, rather than its occupant, adrift).

At the same time that Nieuwenhuis was working on his *New Babylon* project, the abandoned offshore Red Sands Fort and Shivery Sands Tower – both built by the British army for surveillance during the Second World War – were commandeered by pirate radio stations for a period of three years. During this time several legendary pirate radio stations, including Radio City, Radio 390, and Radio Sutch, broadcast from the forts, structurally identical to contemporary oil rigs, in the international waters of the outer Thames Estuary (Cadier, 2002). The brief success of these reoccupations of offshore structures may be as close as the utopian fantasy ever came to realization. Though none of the megastructuralist projects were ever fully realized, the drawings they produced look like optimistic predictions of the elaborate offshore structures being built by the oil industry today. Ironically, the contemporary rig fits Wilcoxon's structural definition of the megastructure perfectly, while the economic and political reality of the rig could not be more antithetical to the utopian ideals around which it was defined.

An equally relevant architectural parallel comes from the work of the Japanese Metabolist architect Kiyonuri Kikutake. Kikutake proposed a mobile, self-propagating *Marine City*. Kikutake's proposal combines the optimistic ideas of the European avant-garde with a decidedly dire forecast of an over-polluted, over-populated future. This forecast overlaps with the largely American science fiction trope in which mankind's gradual corruption and pollution of the earth drive him back into the ocean. This idea, put forth in countless films and novels, proposes that, forced to flee the land (to devolve), man colonizes the ocean with an elaborate network of semi-amphibious metropolises in the sea. Perhaps the best example of this is the 1995 film *Waterworld,* which goes as far as to evoke infamous (if undeniably real) *Exxon Valdez* captain Joseph Hazelwood as the patron saint of the bad guys. It is somewhere between the futurist imaginations of the architectural avant-garde and these dark forecasts that the reality of the rig – the unintended cousin of megastructuralist utopia – seems to lie. In the best-cast scenario, perhaps the reputation of the rig as the icon of the power industry can be reappropriated as the model for a post-terrestrial occupation of the sea? And perhaps the hundreds of decommissioned rigs that have been abandoned across the Gulf of Mexico, the Persian Gulf, the North Sea, and the Campos Basin can be reinvented and reused in the future?

If the idea of mankind retreating to the sea seems far-fetched, it is modest in comparison to an apparently renewed (and politically motivated) interest in expanding spaceward. In so many ways, the rig is the industrial age precedent for space age architecture: the low-technology proto-space station, the machine that enabled colonization where buildings would not do. At once a habitat, a vessel, a machine, the rig is closer to a space station than anything ever built on land and will most likely be seen, some day, as the precursor of the autonomous free-floating environments of the future. Let us only hope that no one finds oil on Mars.

References

Alvarez, A. (1987) *Offshore: A North Sea Journey*. Paris: Flammarion.

Béguin, F. (2001) "Offshore Oil: The Sixth Continent." *Quaderns #18: Architecture and Energy.*

Cadier, A. (2002) "Seafort Based Radio Stations." Marine-Broadcasters.com. http://www.marine-broadcasters.com/seafort%20based%20radiostations.htm (accessed January 8, 2004).

De Zegher, C. and Wigley, M., editors (2001) *The Activist Drawing: Retracing Situationist Architectures from Constant's New Babylon to Beyond.* New York: The Drawing Center.

Deleuze, G. and Guattari, F. (1987) *A Thousand Plateaus: Capitalism and Schizophrenia* (B. Massumi, Trans.). Minneapolis: University of Minnesota Press.

Eaton, R. (2001) *Ideal Cities: Utopianism and the (Un)Built Environment*. London: Thames and Hudson.

Fitzpatrick, C. (1998) "Legal Issues of Ocean Cities" (June). http://www.2100.org/w_oceancitieslegal.html (accessed March 25, 2003).

Foucault, M. (1986) "Of Other Spaces." *Diacritics 16* (Spring), pp. 22–27.

Isaacs, D. (2001) "Rig Hostages Freed in Nigeria." *BBC News* (August). www.bbc.co.uk/1/hi/world/africa/1511857.stm (accessed May 2, 2003).

Kikutake, K. (1959) "Marine Cities." *Kokusai Kenchiku* (January).

Melville, H. [1851] (1972) *Moby-Dick* (or *The Whale*). New York: Penguin.

Sadler, S. (1998) *The Situationist City.* Cambridge, Mass.: The MIT Press.

Saunders, P. (2002) "How the Ocean Ranger Disaster Changed Oil Exploration." CBC News Online (February). www.cbc.ca/news.features/oil_rigs.html (accessed January 10, 2004).

Sekula, A. (2002) *Fish Story*. Düsseldorf: Richter Verlag.

Silverstein, K. (2001) "A Crude Likeness – Oil Tanker Named for National Security Adviser Condoleezza Rice." *Harper's.*

Smith, M. (2003) "Pirate and Offshore Radio." *MDS975.* http://dspace.dial.pipex.com/town/pipexdsl/r/arar93/mds975/content/pirateradio.html (accessed January 8, 2004).

Timmons, H. (2003) "Aging Oil Rigs Raise Safety Issues." *New York Times* (December 30).

Wilcoxon, R. (1968) "Council of Planning Librarians Exchange Bibliography." In R. Banham (1976) *Megastructure: Urban Futures of the Recent Past*. London.

Online references

AFROL News (2002) "Hostage Drama on Nigerian Oil Rig" (April 25). http://www.afrol.com/News2002/nig024_rig_hostages.htm (accessed April 28, 2003).

BBC News Online Network (1998) "Brent Spar Gets Chop" (November 25). http://news.bbc.co.uk/1/hi/world/europe/221508.stm (accessed April 28, 2003).

BBC News Online Network (1999) "Leaks prompt 'Piper Alpha' fears" (September 8). http://news.bbc.co.uk/1/hi/scotland/441520.stm (accessed May 24, 2003).

CBSNews.com (2003) "Details Murky In Oil Rig Hostage Crisis" (April 20). http://www.cbsnews.com/stories/2003/05/02/world552075.shtml (accessed January 2, 2004).

CNN.com/World. (2001) "Major Oil Industry Accidents" March 20) www.cnn.com/2001/world/americas/03/02/oil.accidents/ (accessed April 28, 2003).

Dockwise News Desk (1999) "Dockwise Sets New Record" (December 22). http://www.dockwise.com/news/press221299.html (accessed January 2, 2004).

Forbes.com (2001) "Disaster of the Day: Petrobras" (March 19). www.forbes.com/2001/03/19/0319disaster_print.html (accessed May 2, 2003).

Greenpeace Archive (1996) "Greenpeace Brent Spar Protest in the North Sea" (May 22). http://archive.greenpeace.org/comms/brent/brent.html (accessed May 5, 2003).

Offshore Technology: The Website for the Oil and Gas Industry (2001) "Petrobras Oil Rig and Gas Field Project: Roncador Campos Basin, Brazil" http://www.offshore-technology.com/projectID=1788 (accessed April 28, 2003).

"The Lay of the Land": The Center for Land Use Interpretation Newsletter (1997) "Unique Commercial Satellite Launch Complex Under Construction in Long Beach" (Summer). www.clui.org/clui_4_1/lot1/lot1v11/sealaun.html (accessed November 28, 2003).

A Generation on the Move: The Emancipatory Function of Architecture in the Radical Avant-garde 1960–1972

Renata Hejduk
Arizona State University

> The future environment will be where you (yourself) may find it. If there is an upsurge in designing for mobility of society and mobility of facilities, so that the previous limitations of location and institution are overridden, we shall reach a point where the whole of territory is part of a responsive environment.
> (Cook, 1970)

This chapter provides a brief introduction to the cultural and philosophical framework that preceded and initiated the contemporary discussion about architectural environments that move, expand, plug in, unplug, tune in, and turn on. Although this story ultimately has its roots in the early twentieth century, the work of a number of British and European architects and urbanists (often called the radical avant-garde) will be discussed within the greater intellectual and cultural milieu of bodily and psychological freedom, liberation, and permissiveness that exploded during the 1960s and early 1970s (Figure 5.1). Themes of liberation and freedom were commonplace by the mid-1960s: the civil rights, feminist, and free-speech movements were only a few of the numerous peaceful (and not so peaceful) coalitions that formed under the banner of freedom. A dominant theme, characteristic, or feature that emerges upon close examination and comparison of these designers is the idea of an architecture that is itself mobile or encourages mobility and freedom of the mind and body within its structures. The works discussed here are mainly theoretical, prototypical, or conceptual in nature.

5.1 Dancer at a love-in, Golden Gate Park, 1968.
Photo by: earl leaf/michael ochs archives.com

The main argument of this chapter, regarding the concept of freedom and emancipation as produced in the work of the radical avant-garde, is contingent upon the secondary argument that there is an identifiable transnational culture of liberation and freedom. This is often described as one of the salient features of the 1960s, and much of the work of the emerging radical architects of this period deals with the notion of both physical and psychological "liberation" aided by the postwar advancements in technology and materials.

Introduction

Herbert Marcuse and Norman O. Brown were both seen as major social theorists who helped define the counterculture that was to emerge in the late 1950s and throughout the 1960s in the United States and abroad. Herbert Marcuse's (1955) *Eros and Civilization: A Philosophical Inquiry into Freud* was a discussion of the implications of Sigmund Freud's *Civilization and its Discontents*, its thesis regarding civilization and its repressions. Marcuse, whose work was well known in intellectual European circles, became internationally famous with the publication of *Eros and Civilization* which was a key work in the intellectual legacy of the 1950s, and very important in shaping the new subcultures of the 1960s. His work was translated by the influential French literary/theoretical magazine *Arguments* during the late 1950s and 1960s, and also was widely translated and published in student journals. Within three years of *Eros and Civilization*'s publication, Norman O. Brown published *Life Against*

Death: The Psychoanalytic Meaning of History (1958), which quickly became an underground classic. Although not as easily or widely read as *Eros and Civilization*, Brown's maverick yet influential ideas were often cited by both the early participants of the counterculture (such as Abbie Hoffman in the United States) and the later generation that followed them, those who came of age in the late 1960s. Both books were to encourage the liberal pulse that was beginning to beat on both sides of the Atlantic. Although their methodologies and political positions are quite different, Marcuse and Brown did agree upon one primary point: that to effect a social change man's consciousness had to be awakened or demystified and his repressions freed.

Marcuse's critique of technology was extremely influential on the "New Left." He asked: "Is it still necessary to repeat that science and technology are the great vehicles of liberation, and that it is only their use and restriction in the repressive society which makes them into vehicles of domination?" (Marcuse, 1969, p. 12). As a follower of Marx, Marcuse found modern man alienated from his life by his relationship to the mode of production and the invisible structures that encapsulated and defined his life. *Eros and Civilization* sought to synthesize a socialist notion of economic revolution with a "more anarchistic and utopian idea of emotional revolution" (Cranston, 1979). According to Marcuse, with the development of the forces of production such as mechanization and automation, the "historical necessity for existent forms of repression is undermined. . . . Automation promises the end of the use of the body as a mere instrument of production. The technical need for sexual repression can be challenged." (Rosak, 1968, pp. 122–125) Freud argued that for civilization to flourish the individual had to repress his personal desire for the good of the group. According to Marcuse, instead of harboring and repressing our desire and pleasure, technology had the capacity to free and motivate desire.

Marcuse's *Eros and Civilization* and *One-Dimensional Man* (1964), and Brown's *Life Against Death* were exceptionally influential in the freedom and liberation movement. The counterculture used their theories along with those of Wilhelm Reich, R. D. Laing, Marshall McLuhan, and Timothy Leary to help produce their rhetoric of freedom. The thrust of their argument was that civil society imposed restraint and restriction on the freedom of the individual. "Basically, it's just a question of freedom. It's your body – you can do with it what you want to" (Gross, 1968, pp. 61–62). It was up to the individual to counter this societal restraint and produce a more playful and expressive life. This newly found freedom was expressed in many forms. The hippies renounced the conservative taboos against physical contact and they practiced sexual liberation; Tantric Buddhism, Freud, Brown, and Reich inspired them, and they condoned homosexual relationships. The fruits of this growing radical culture of liberation and freedom could clearly be seen as manifested in everyday life. Examples of this would be the acceptance of profanity in everyday language, sartorial freedoms such as blue jeans replacing pressed trousers as accepted daily wear, and the new subjects that emerged from the nascent freedom movements. The liberated woman, the liberated youth, and the energized civil rights movement were all expressions of this enormous shift in society.

The critique of the Modern Movement

An underlying assumption here was that the rationalism, functionalism, and mode of production espoused by the Modern Movement was analyzed and critiqued by this third generation of architects who ultimately found the "Modern project" to be unfinished, alienating, and repressive. They looked to the concept of freedom (and an architecture and urbanism that explored and celebrated it) to resolve those conflicts that arose from the modern condition. It was the modern ideal of architecture as rational and functional, and its utopian desire to engender a better world through the symbolic representation of the mode of production, that this generation of architects who grew up during the war both celebrated and critiqued. This generation of young European and British architects who went to school during the 1950s and began to practice and teach during the late 1950s and early 1960s were less convinced by Modernism's social and formal aims, but were attracted to the architectural potential of advances in technology that were the handmaiden of progress. Instead of looking to the more formalized work of Le Corbusier and Mies van der Rohe, they championed designers such as Buckminster Fuller whose Dymaxion House and transportable/mobile structures held promise and inspiration for their own emerging practices. Groups such as Archigram, Haus-Rucker, and Superstudio, and architects such as Hans Hollein among many others embraced the ideology of liberation, freedom, and pleasure that permeated Europe and the United States during the 1960s. Archigram member Warren Chalk's optimistic technological rhetoric described their aims and clearly illustrated their knowledge of counterculture gurus such as Marshall McLuhan.

> The world of architecture will eventually move away from the idea of buildings as something fixed, monumental, great and edifying, into a situation where buildings take their rightful place among the hardware of the world. Then architects as presently known will cease to exist, and a very different kind of animal will emerge, embracing science, art and technology in a complex overview. Established disciplinary boundaries will be removed and we will come closer to the all-at-once world of Marshall McLuhan.
> (Chalk, 1966a, pp. 172–173)

They began to exploit technology and rationality in architecture with the hope of counteracting the societal and bodily repressions of late capitalism. For groups such as these, the ideology of technology in the modern condition had not been fully explored and had resulted in fairly disastrous effects. Norman O. Brown mythically recounted the effects of the modern condition in the final chapter of *Life Against Death*. He took the following quote from *Sunday after the War*, a story written in 1944 by Henry Miller; it is in the introduction to Part 6 entitled "The Way Out." In this passage Miller mused on the world after war, the new civilization:

> The cultural era is past. The new civilization, which may take centuries or a few thousand years to usher in, will not be another civilization – it will be the open stretch of realization which all the past civilizations have pointed to. The city, which was the birthplace of civilization, such as we know it to be, will exist no more. There will be nuclei of course, but they will be mobile and fluid. The peoples of the earth will no longer be shut off from one another within states but will flow freely over the surface of the earth and intermingle. There will be no fixed constellations of human aggregates. . . . The machine will never be dominated, as some imagine; it will be scrapped, eventually, but not before men have understood the nature of the mystery which binds them to their creation. . . . Man will be forced to realize that power must be kept open, fluid, and free. His aim will be not to possess power but to radiate it.
> (quoted in Brown, 1958, p. 305)

The language and the images that he provided were practically exact models for the work of the radical avant-garde. The image of people roaming freely over the surface of the earth, across landscapes that were no longer bounded by the idea of states or of fixed cities, could be the exact description of the work of any of the first generation of radical architects such as Archigram, Utopie, Archizoom, and Superstudio. It was especially evocative of Superstudio's *Supersurface Five Fundamental Acts: Life, Education, Ceremony, Love and Death* (1972–73), in which the landscape was hardwired to allow us to create our own environment in whatever natural setting that we chose (Figure 5.2). The work of these groups clearly expressed the designed manifestation of Marcuse's attempt to find the potential in technology, Norman O. Brown's utopian hope for the freedom of the body, the general culture of mind and body, and the place of technology in the 1960s. In addition, the idea of a rationalized grid of infrastructure (often invisible) that freed communication, circulation, and mobility was the endgame to a functionalism and rationalism that were seen as not having gone far enough. The problem with modernism was not its ideology, but its unwillingness to take that ideology to its ultimate conclusion. As we will see, the radical avant-garde of the 1960s and early 1970s illustrates how the commitment to a critique of modernism and a continued investigation of the potentials of technology can liberate both the individual and society from the repressive binds of an architecture unable to shed its ideological restraints.

Technology, mobility, and liberation

> In this century there have been several occasions when science, technology, and human emancipation have coincided in a way that has caused architecture to explode.
> (Cook, 1970, p. 11)

> Is it still necessary to repeat that science and technology are the great vehicles of liberation, and that it is only their use and restriction in the repressive society which makes them into vehicles of domination?
> (Marcuse, 1969, p. 12)

As Europe recovered from the ravages of World War II, many of the youth of Great Britain, France, and Italy turned away from the brute reality of their broken cities and shattered lifestyle and began to seek a new life through education. The British government set up numerous programs to allow for the training and retraining of their student-age populations. The situation was not as proactive in countries such as Austria where the educational system in architecture retained the strictures of *beaux-arts* training, but nevertheless the Austrian students of the 1950s and early 1960s managed to produce some of the most experimental and provocative work of the period. As stated earlier, following the war years many of the young designers were unconvinced by the strict tenets of modernism which they felt had, in many ways, created a sterile and socially unresponsive urban condition. The Archigram group, the Italian radical avant-garde groups Superstudio and Archizoom, and the early work of the Austrians Hans Hollein, Haus-Rucker, and Coop Himmelblau emerged from this milieu and engaged the culture of liberation and the hope of technology to create an architectural image of this philosophy/ideology.

In late 1950s, a group of young British men began to discuss the changing scene of art and architecture that they saw forming in London. They wanted to continue the polemic of architecture school (the discussion and the critique), and they realized that a publication might help to bring their ideas to the fore and engender a critical mass. They decided to self-publish a magazine called *Archigram*, as thin as a comic book, that would be easy to ship and quick to consume. In projects such as *Instant City* and *Airhab*, Archigram illustrated their desire to produce works of architecture that responded to situations, instead of definite, defined, immovable structures that resisted permutation (Figure 5.3). For them, in a Marcusian vein, architecture had to use and explore advanced technology and through its use architecture would become programmatically more flexible: it would react to the changing needs and desires of its users. Archigram member Warren Chalk states: "In a technological society more people will play an active part in determining their own individual environment, in self determining a way of life" (Chalk, 1963, p. 92). Their

5.2 Superstudio, *Supersurface Five Fundamental Acts*, 1972. © Cristiano Toraldo di Francia

5.3 Peter Cook, *Instant City*: typical configuration, highway at the intersection of Santa Monica and San Diego, Los Angeles, 1969. Photo: Christian Wachter, Kunsthalle, Vienna

works became increasingly nomadic and portable as the decade progressed. In their writings, their language was infused with ideas that equated a new architectural form with psychological and behavioral freedom. The specialization that technology had engendered thus far led man to a "skillful but spiritless existence; people with enormous fatigue trying to cope with the banalities of not-too-well-serviced environments" (Chalk, 1963, p. 92). Their technologically promiscuous architecture and city forms would enable the inhabitant to be more expansive and break through boundaries of psychic and bodily repression and would lead them towards a liberating new production of life.

As stated by Peter Cook, "[w]e are not politically over-developed as a group, but there is a kind of central emancipatory drive behind most of our schemes" (Cook, 1967). The ultimate aim for Archigram was the emancipatory function of architecture. In an almost Freudian exercise of free association, Warren Chalk extolled the virtues of this new society enabled by technological freedom. The new model for the individual is a "technological opportunist – an inventor." The inventor will be "a breaker of boundaries. . . . New associations (soft solutions) allow us to get closer to the ultimate pleasure of being through their non-binding means of arrangement" (Chalk, 1963). Dennis Crompton writes in *Archigram 8*: "If the environmental business is concerned with the extension of man's experience then the means of achieving this is by pushing current technology" (Crompton, 1968, p. 257). The new man or woman would look not to connections that held things together like glue, but to a "less physical solution." An example of this idea of sensory pleasure through a flexible design object would be Mike Webb's *Cushicle* that premiered in *Archigram 7* (1966) (Figure 5.4).

5.4 Archigram, *Cushicle*, Maquette, 1966, Photo: Christian Wachter, Kunsthalle, Vienna

> The Cushicle is an invention that enables man to carry a complete environment on his back. It inflates-out when needed. It is a complete nomadic unit – and it is fully serviced. It enables an explorer, wanderer or other itinerant to have a high standard of comfort with a minimum of effort. The Cushicle carries food, water supply, radio, miniature projection television and heating apparatus. The radio, TV, etc. are contained in the helmet and the food and water supply are carried in pod attachments. With the establishment of service nodules and additional apparatus, the autonomous Cushicle unit could develop to become part of a more widespread urban system of personalized enclosures.

Archigram 7 was a watershed issue for the group. The magazine contained numerous projects that mobilized their already inventive and radical architectural propositions. Peter Cook's *Blow-out Village* presented a transportable environment that could be used in disaster areas or for less serious and more playful events. "Mobile villages [could] be used everywhere to rehouse people hit by disaster, for workmen in remote areas, and as fun resorts sited permanently or seasonally at the seaside and near festivals. When not in use the village [was] quarter size." *Archigram 7* (and the work done between 1966 and 1967) codified the group's interest in soft and mobile solutions for the critique of dwelling and permanence. Michael Webb's *Suitaloon* borrowed the idea of the space suit from NASA – the space suit could be identified as a minimal house – and reinvented it as a housing solution. Its pseudo-byline reads: "Clothing for living in – or, if it wasn't for my Suitaloon I would have to buy a house." The *Suitaloon* allowed one complete mobility and protection while roaming through the environment:

> Each suit has a plug serving a similar function to the key to your front door. You can plug into your friend and you will both be in one envelope, or you can plug into any envelope, stepping out of your suit which is left clipped on to the outside ready to step into when you leave. The plug also serves as a means of connection envelopes together to form larger spaces.

In 1968 the whole group was asked to exhibit at the Milan Triennale. By marrying Webb's *Suitaloon* with Cook's *Blow-out Village* they proposed the *Inflatable Suit-Home* (Figure 5.5). Their statement for the Triennale reads:

> The ability of objects and assemblies to metamorphose over a period of time so that we are no longer stuck with monuments of a forgotten day . . . the ability to use the world's surface and mobility to achieve personal freedom: The nomadic instinct and the nomadic potential of cars and car based enclosures . . . the realization that although we are beginning to be emancipated socially, economically and through a consumer society, building has not caught up with this range . . . the interplay of man and machine to develop this responsive environment and the free ranging exchange of all as and when needed . . .
> (Cook *et al.*, 1999, p. 83)

Along the same line of thought, Archigram member David Greene surmises: "More and more people want to determine their own parameters of behaviour. They want to decide how they shall be, whether it's playing, working, loving, etc." Greene includes a fragment of a poem:

> I have a desire for
> The built environment
> To allow me to do
> My own thing
> (Cook *et al.*, 1966, p. 306)

Interest in portable minimal dwelling was not an isolated interest within the British avant-garde. In 1968, a year after Peter Cook published his *Blow-out Village*, Antoine Stinco of the French Utopie Group debuted his diploma project for the École Nationale Supérieure des Beaux-Arts. It was called "Itinerant Exhibition Hall for Objects of Everyday Life" and proposed a hall for the exhibition of mundane objects that could be instantly created from the arrival of several trucks containing the inflatable components. The project seemed consciously or subconsciously inspired by and responding to Archigram's call for instant cities and buildings, as well as Frei Otto's work with pneumatic and tensile structures. As Marc Dessauce illustrated in his excellent history of this design movement, *The Inflatable Moment: Pneumatics and Protest in '68* (1968), Utopie was among a crowd of young avant-gardes fully invested in the promise of pneumatics as a response to the alienation of functionalism and rationalism. This was illustrated in 1968, when the newly formed Austrian team of Haus Rucker (Ortner, Pinter, and Zamp) debuted two pneumatic projects: *Pneumatic Space for Two People* (also known as *The Mind Expander*) and *Yellow Heart* (Figure 5.6). The inflatable mobile 'movement' continued to gather strength in 1969 when Hans Hollein inflated his *Mobile Office* and went to work (Figure 5.7). In 1971, Coop-Himmelblau inflated their *Restless Ball* (Figure 5.8) and rolled through the streets of Vienna, and in 1972 Haus Rucker debuted *Oasis Number 7* at Documenta #5 (Figure 5.9).

In an article entitled "Alles ist Architecture" for *Architectural Design* (1970), Hans Hollein recounted how the students in Vienna and Graz were stifled under the Austrian system that tried to repress any information about avant-garde work that was being done elsewhere. Some professors even forbade the students to see the exhibition *Architektur* by Hans Hollein and Walter Pichler that was on view for only four days. The students disregarded this form of censorship and were inspired and rallied by the radical exhibition. Hollein went on to state that much of this work could be seen in relation to the strong Freud-based Austrian heritage of operating psychologically on the individual user's mind and consciousness as well as on the immediate environment while introducing ideas of individuality, fun, relaxation, leisure-time activities, and action. At the same time their desires were encouraged by

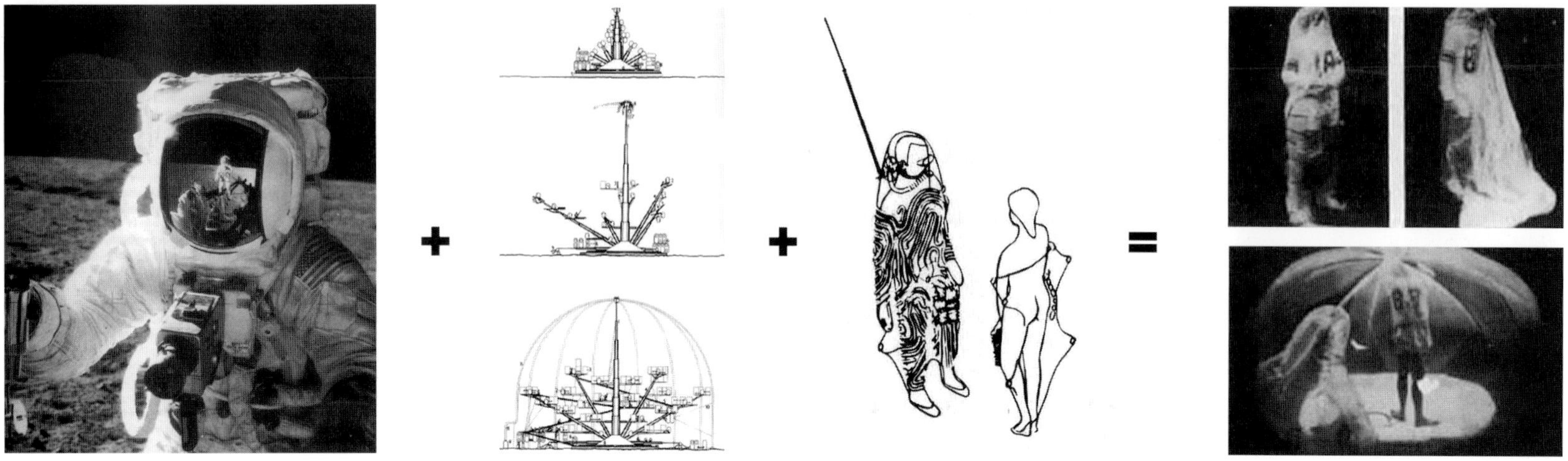

5.5 NASA Space Suit 1 Archigram, *Blow-out Village*, 1966, 1 Suitaloon,1966 5 *Inflatable Suit Home*, 1968, Photo Archives Archigram

5.6 Haus-Rucker, *Mind Expander and Yellow Heart*, 1968. © Ortner and Ortner

5.7 Hans Hollein, *Mobile Office*, 1968. © Hans Hollein

5.8 Coop-Himmelblau, *Restless Ball*, 1971. © Coop-Himmelblau

5.9 Haus-Rucker, *Oasis Number 7*, 1972. © Ortner and Ortner

information about the work that was being done by groups such as Archigram which were especially popular among the students (Hollein, 1970, pp. 60–63).

In 1972, the work of individuals such as Ettore Sottsass, Alberto Rosselli, Marco Zanuso, and Richard Sapper, and of Superstudio and Archizoom among others, was introduced to the American design scene in a ground-breaking museum exhibition, entitled *Italy: The New Domestic Landscape: Achievements and Problems of Italian Design,* curated by Emilio Ambasz for the Museum of Modern Art (MOMA), New York (see Ambasz, 1972). This was the first time that many American designers had seen the work of the Italian radical avant-garde and the show was a watershed moment. It was clear that the radical culture and modernist critique of the 1960s had permeated deeply into the avant-garde psyche. The Italian radical avant-garde took the critique of Modernism to an extreme and politicized level. They were not satisfied with Archigram's criticism that Modernism had missed the point about technology and progress. They wanted architecture to emancipate the human subject from the binds of functionalist repression to reveal the contradictions and ambiguities at the basis of the Modern Movement, and to deal with the fundamental material, rites, and actions of our human existence. They wanted architecture to emancipate people from *architecture*.

According to the Italian avant-garde, the problem with Modernism was not its ideology; it was its unwillingness to take that ideology to its ultimate conclusion, and also its commitment to architecture that formally resembled technology. They even critiqued the work of Archigram and the Japanese Metabolists for being an image of technology, i.e. it looked like technology. For Archizoom Associati, to pretend that functions had particular shapes was a false qualification of Modernism that would be eliminated. The single plane of the skyscraper and housing unit, the supermarket, and the grid would be exalted and imposed, instead of being implied and hidden within a rhetoric of a relationship with nature. The factory and the supermarket become the specimen models of the future city: optimal urban structures, potentially limitless, where human functions were arranged spontaneously in a free field, and made uniform by a system of micro-acclimatization and optimal circulation of information. The natural and spontaneous "balance of light and air is superseded: the house becomes a well-equipped parking lot. Inside it there exist no hierarchies or spatial figurations of a conditioning nature" (Branzi, 1992, p. 50) (Figure 5.10). Archizoom's *No-Stop City* was to be made up of a totally artificial and homogenous infrastructure of residential districts. These districts would contain large micro-acclimatized and artificially illuminated spatial systems, much like the Home Depot or Wal-Mart in the United States, that were available for undifferentiated and continuous use. These free and equipped spatial systems were called "Residential Parking" lots that would allow inhabitants to appropriate for their individual and spontaneous uses.

The rhetoric of an environment that allowed one to do one's own thing freed from the encumbrances of functional architectural forms was especially apparent in the Florentine group Superstudio's *Supersurface, Five Fundamental Acts: Life, Education, Ceremony, Love and Death (1971–73)* as published in the Italian design magazine *Casabella*. The following is the description from the MOMA exhibition catalogue (Superstudio, 1972, pp. 240–251):

> We can imagine a network of energy and information extending to every properly inhabitable area. Life without work and a new "potentialized" humanity are made possible by such a network. The network of energy can assume different forms. . . . It is an image of humanity wandering, playing, sleeping, etc. on this platform. Nomadism becomes the permanent

5.10 Archizoom Associati, *No-Stop City*: Internal Landscape, 1970. © Andrea Branzi

condition. . . . There will be no further need for cities or castles. There will be no further reason for roads or squares. Every point will be the same as any other. So, having chosen a random point on the map, we'll be able to say my house will be here for three days two months or ten years. . . .You can be where you like, taking with you the tribe or family. There's no need for shelters, since the climatic conditions and the body mechanisms of thermoregulation have been modified to guarantee total comfort. At the most we can play at making shelter, or rather at the home, at architecture . . . .

The work of Superstudio and Archizoom were the more extreme and anti-architectural manifestations of the Italian radical avant-garde's desire for freedom from the current fixed architectural situation. On a more pragmatic level were the exhibition entries of Ettore Sottsass, Joe Columbo, and Alberto Rosselli; all presented mobile dwelling units "complete and fully equipped habitations, easily transportable and ready for immediate use." In Alberto Rosselli's description of his *Mobile House* (Figure 5.11) that was exhibited at MOMA, he used descriptive language that recalled Archigram's quest for new technologies and materials to help transform the idea of living with and within architecture.

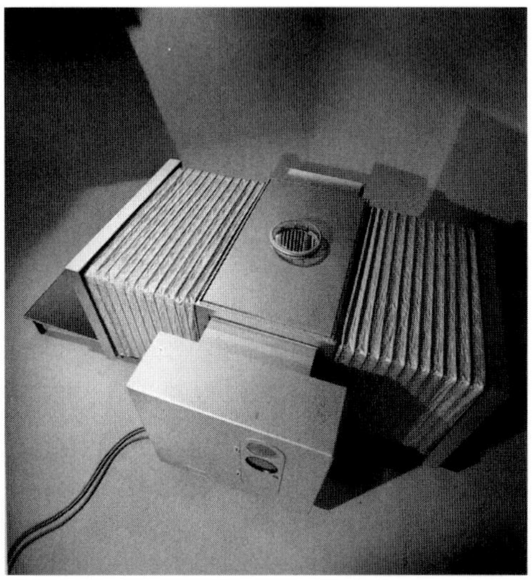

5.11 Alberto Rosselli, *Mobile House*, 1972

> Contemporary technology permits us to extend mobility and expansion through the use of lightweight materials and more highly developed mechanisms. . . . It is indeed possible to envisage a house that conforms to the psychological requirements of life, an object that can be transformed according to the various uses to which it will be put, and that after a certain time can be completely reassembled.
> (Rosselli, 1972, p. 182)

For these designers, the mobile dwelling units that they proposed for the exhibition were not only basically equipped to meet the minimum functions of daily life, but also there was an inherent ideological/political rhetoric apparent in almost all their project descriptions (also found in the writings of Archizoom and Superstudio). These prefabricated, manufactured mobile dwelling units and micro-acclimatized infrastructures didn't merely give us the freedom to roam the earth like a band of nomadic hippies. These objects and situations, presented as a set of ideas, encourage us to minimize our dependence on the accumulation of material objects, and on the obsolete idea that the functions and rites of life are dependent upon codified forms of space. This mindset that they challenged at the time is akin to Marcuse's idea of one-dimensionality where the structures of the marketplace, capitalism, and the division of labor are so sublimated into our everyday existence that they appear natural. Norman O. Brown writes:

> [t]he resurrection of the body is a social project facing mankind as a whole, and it will become a practical political problem when the statesmen of the world are called upon to deliver happiness instead of power. . . . Contemporary social theory has been completely taken in by the inhuman abstractions of the path of sublimation, and has no contact with concrete human beings, with their concrete bodies, their concrete though repressed desires.
> (Brown, 1958, pp. 317–318)

The potential of mobile and transportable environments proposed by the "Italian Radicals" is that they dialectically both draw from and critique the ideals of Modernism and modernity while at the same time offering more concrete (albeit theoretical and utopian) solutions to the questions posed by the liberated philosophy of the era. "An idea like this can only work on the assumption that the 'rites' of life . . . can begin a new morning with a new awareness of existence" (Sottsass, 1972, pp. 162–163). Ultimately, their work promotes physical and psychological freedom as its central theme and ideological critique as its methodology.

In his book *Experimental Architecture* (1970), Peter Cook surveys the previous decade's cutting-edge work and notes that

> [l]inked with these general advances in technology is an entirely new architectural concept: that man can have his own container. This suggests that each person, on arriving at a state of relative emancipation, should receive a degree of personal support that he cannot get from the collective artifact.

He presaged that "life-style" would be prompted increasingly from experience outside home and that new demands would be made upon houses, office buildings, and schools, and that they would have to adapt to different psychological stimuli. For Cook, man's liberation would be produced by the interface of his desires with technology. Machines and technology were beginning to respond to "our psychological demands" and he predicted "a true symbiosis of the person and the artifact" (Cook, 1970, pp. 133–152). No longer would architecture dictate the form that functions would take, but the human mind/body and its ultimate freedom would be instrumental in dictating the future of architecture.

The emancipatory projects of the European and British radical avant-garde during this period were a direct result of the climate and narrative of liberation and freedom that bloomed during the 1960s. The solutions they proposed ranged from invisible infrastructures that enabled mass human exodus across a freshly wired landscape to miniaturized mobile dwelling units that delivered the minimum needs for human habitation. The ultimate pleasure of being and the rites of life were dependent upon an altruistic technology. Other issues that need to be addressed relative to this discussion involve an understanding of the nature of the human rights movement, initiated after the atrocities of World War II, and the impact of that ideal of freedom upon this generation after the postwar period. Additionally, the recent surge in architectural projects that are mobile and transportable need to be addressed relative to the current political and cultural climate in the United States and abroad. It is hoped that this brief introductory chapter lays the groundwork for a more deliberate discussion on the impetus for and implications of the contemporary trends in mobile architecture.

References

Ambasz, E. (ed.) (1972) *Italy: The New Domestic Landscape. Achievements and Problems of Italian Design.* New York: The Museum of Modern Art in collaboration with Centro Di, Florence.

Banham, R. (1965) "A Home Is Not a House." Originally published in *Art in America* (April 1965). Reprinted in J. Ockman (1993) *Architecture Culture 1943–1968: A Documentary Anthology*. New York: Columbia Books of Architecture/Rizzoli.

Banham, R. (1965) "The Great Gizmo." Originally published in *Industrial Design 12*. Reprinted in Mary Banham *et al.* (1996) *A Critic Writes: Essays by Reyner Banham*. Berkeley: University of California Press, pp. 120–121.

Bokin, J. and Lukes, T. (1994) *Marcuse from the New Left to the Next Left*. Lawrence: University of Kansas.

Branzi, A. (1992) *Andrea Branzi: The Complete Works*, London: Thames and Hudson.

Brown, N. (1958) *Life Against Death: The Psychoanalytic Meaning of History*. Wesleyan: Wesleyan University Press.

Chalk, W. (1963) "Housing as a Consumer Product." Reprinted in T. Stoos (ed.) (1994) *A Guide to Archigram: 1961–74*. London: Academy Editions.

Chalk, W. (1966a) "Hardware of a New World." Originally published in *Forum*, October. Reprinted in T. Stoos (ed.) (1994) *A Guide to Archigram: 1961–74*. London: Academy Editions.

Chalk, W. (1966b) "The 40's." Originally published in *Archigram* 6. Reprinted in T. Stoos (ed.) (1994) *A Guide to Archigram: 1961–74*. London: Academy Editions.

Chalk, W. (1971) "Touch Not." Originally published in

Architectural Design, April. Reprinted in T. Stoos, (ed.) (1994) *A Guide to Archigram: 1961–74*. London: Academy Editions, pp.358–359.

Cmiel, K. (1994) "The Politics of Civility," in D. Farber (ed.) (1994) *The Sixties: From Memory to History*. Chapel Hill: University of North Carolina Press.

Cook, P. (1966) "Blow-out Village." *Archigram 6*. Reprinted in T. Stoos (ed.) (1994) *A Guide to Archigram: 1961–74.* London: Academy Editions.

Cook, P. (1967) "Some Notes on the Archigram Syndrome." *Perspecta*, Supplement no. 11, Yale University.

Cook, P. (1970) *Experimental Architecture*. New York: Universe Books.

Cook, P. *et al*. [1972] (1999) *Archigram*. Originally published: Basel and Boston: Birkhauser. Re-published: New York: Princeton Architectural Press.

Cranston, M. (1979) Obituary of Herbert Marcuse. *Guardian*, July 31.

Crompton, D. (1968) "The Piped Environment." Originally published in *Archigram 8*. Reprinted in T. Stoos (ed.) (1994) *A Guide to Archigram: 1961–74*. London: Academy Editions.

Dessauce, M. (1999) *The Inflatable Moment: Pneumatics and Protest in '68*. New York: Princeton Architectural Press.

Farber, D. (1994) *The Sixties: From Memory to History*. Chapel Hill: University of North Carolina Press.

Feenberg, A. (1994) "The Critique of Technology: From Dystopia to Interaction." in J. Bokin and T. Lukes (eds) *Marcuse from the New Left to the Next Left*. Lawrence: University of Kansas.

Greene, D. (1970) "Lawun." Originally published in *Archigram 9*. Reprinted in T. Stoos (ed.) (1994) *A Guide to Archigram: 1961–74*. London: Academy Editions, pp. 306–307.

Gross, H. (1968) *Flower People.* New York: Ballantine Books.

Hollein, H. (1970) "Alles ist Architecture." *Architectural Design*, February, pp. 60–63.

Jameson, F. (1971) NJ *Marxism and Form: Twentieth Century Dialectical Theories of Literature*. Princeton, NJ: Princeton University Press.

Lang, P. and Menking, W. (2003) *Superstudio: Life without Objects*. Milan: Skira Editore S.p.A.

Lukes, Timothy J. (1994) "Mechanical Reproduction in the Age of Art: Marcuse and the Aesthetic Reduction of Technology," in J. Bokin and T. Lukes (eds) *Marcuse from the New Left to the Next Left*. Lawrence: University of Kansas.

Marcuse, H. (1955) *Eros and Civilization: A Philosophical Inquiry into Freud*. Boston: The Beacon Press.

Marcuse, H. (1964) *One-Dimensional Man: Studies in the Ideology of Advanced Industrial Society.* Boston: Beacon Press.

Marcuse, H. (1968) "Liberation from Affluent Society," in D. Cooper (ed.) *To Free a Generation: The Dialects of Liberation.* New York: Macmillan.

Marcuse, H. (1969) *Essay on Liberation*. Boston: The Beacon Press.

Marwick, A. (1998) *The Sixties: Cultural Revolution in Britain, France, Italy and the U.S. 1958–1974*. New York: Oxford University Press.

Ockman, J. (1993) *Architecture Culture 1943–1968: A Documentary Anthology.* New York: Columbia Books of Architecture/Rizzoli.

Pettena, G. (1982) *Superstudio: 1966–1982. Storie, Figure, Architettura.* Florence: Electa Firenze.

Poster, M. (1975) *Existential Marxism in Postwar France: From Sartre to Althusser.* Princeton, NJ: Princeton University Press.

Rosak, T. (1968) *The Making of a Counter Culture: Reflections on the Technocratic Society and Its Youthful Opposition*. Berkeley: University of California Press.

Rosselli, A. (1972) "Mobile House." in E. Ambasz (ed.) *Italy: The New Domestic Landscape. Achievements and Problems of Italian Design*. New York: The Museum of Modern Art in collaboration with Centro Di, Florence.

Sottsass, E. (1972) "Project Description." in E. Ambasz (ed.) *Italy: The New Domestic Landscape. Achievements and Problems of Italian Design.* New York: The Museum of Modern Art in collaboration with Centro Di, Florence.

Stoos, T. (ed.) (1994) *A Guide to Archigram: 1961–74.* London: Academy Editions.

Superstudio (1972) "Description of the Microevent/Microenvironment." in E. Ambasz (ed.) *Italy: The New Domestic Landscape. Achievements and Problems of Italian Design*. New York: The Museum of Modern Art in collaboration with Centro Di, Florence.

Unger, I. and D. (1998) *The Times Were a Changin': The Sixties Reader*. New York: Three Rivers Press.

Carried Away! The Spatial Pleasure of Transportability

Patricia Pringle
Royal Melbourne Institute of Technology

> 'Flexibility and mobility lead to transparency, openness, freedom and spatial dynamism.'

This assertion from the young Marcel Breuer (Wilk, 1981, p. 67) is one of the design mantras of the early twentieth century. It seems to speak for a sort of magical thinking, where the naming of one desirable attribute will invoke the next. In this chapter it is used to open a discussion of our human fascination with such dynamic operations as dematerialising, defying gravity, breaking spatial constraints or changing form, which have taken on renewed urgency in the design practices of this new century.

With hindsight it is easy to see that the transparency desired by Breuer was more than the simple opposite of opacity. In their seminal essay 'Transparency: Literal and Phenomenal', Colin Rowe and Robert Slutzky note that in the literature of contemporary architecture words like transparency, space-time, simultaneity, interpenetration, superimposition, ambivalence are often used as synonyms (1997, p. 22). By mid-century, the complex conceptual associations of 'transparency' had come to be taken as axiomatic of spatial modernity.

Today, 'transportability' forms part of a similarly charismatic set of ideas that are assumed, almost without a second thought, to be related. Associated manoeuvres – folding, transforming, inflating, expanding, reconfiguring – are all keywords for entry into a web of creative thought, whether on spatial design, industrial design or fashion, which could form a list of 'synonyms' for today. Pursuing this avenue of exploration might lead to an examination of 'literal and phenomenal transportability' and it might indeed be fruitful to explore the renewed concern with the kinaesthetic experience in architecture. However, I want to avoid the hegemony of architecture in this discussion and explore something more than the notion of spatial experience in our *inner* world. In this chapter I want to consider some ways in which 'transportability' can be a fascinating magical attribute as well as a utilitarian one.

The use of the term 'magical' is a reference not to the supernatural but rather to those fragile human moments when we become aware that we are in a place that is different from the one we normally occupy, where things are connected in a way that is different from what we have assumed. Our attention stems from the object itself and extends to a succession of fleeting thoughts and impressions. At the heart of the matter is always some sense of duality: of oscillation, simultaneity, multiplicity, of shifting (either from here to there, or from one form to another, or from one meaning to another) and this oscillation is in some way satisfying, exciting or profound. We too are transported – perhaps our hearts or our thoughts leap – we may feel a kind of shock or thrill and a quickening of vital emotion that may express itself in laughter, or in wonder, or in a sense of delight at the perception of other possibilities.

There is an essential ambivalence at the heart of the definition of 'the transportable', or how else would we know it? We can only grasp it in relation to the image of the 'non-transportable' against which we are assessing it. Already we have at least two thoughts with which to engage. Like most important words, 'transportable' has multiple meanings that are opposite. At one extreme it may promise to let us carry everything with us, as with the mobile home, while at another extreme it may suggest that we need carry almost nothing, for we can make a place by stretching the wedding canopy, lighting a match or drawing a line in the sand. An idea can be felt to have moved, just as much as an object. There is a play on thought as we probe the ambiguities of this concept.

But what is it about all this that is able to *fascinate* us? Fascination, in its original context of bewitchment, rendered its victim unable to move or resist. In modern usage it carries a sense of 'delighted attraction'. Something is present which is important enough to hold us in a state where we feel connected to it. This goes further than simply seeing it, but holds us on the threshold of fully understanding it.

I have already suggested that today the word 'transportable' has acquired complex connotations for design, just as 'transparency' once did. It embraces the dualities of Rowe and Slutzky's synonyms (space-time, simultaneity, etc.) and further invokes the shape-shifting spatial thoughts that tantalise designers (ambiguity, mobility, ephemerality, lability, etc.). The words, and the 'equivocal sensations' (Rowe and Slutzky, 1997, p. 43) that they conjure up, express our human need for the world to be richly significant – to 'enjoy the sensation of looking through a first plane of significance to others lying behind it' (Rowe and Slutzky, 1997, p. 23).

This concept of 'enjoying the sensation' reinforces the argument that moves the transportable object away from being an end in itself and into 'transportability' as the ability (of the thing or the situation) to transport us, to shift our frame of reference, to permit us, albeit momentarily, to see and feel *creatively*.

Alfred Schutz, the philosopher and sociologist, whose lifework was a concern for the meaningful structure of daily life, offers the idea that each of us experiences reality as many different states, temporarily emigrating from the 'paramount reality of everyday life' to enclaves within it which he called 'finite provinces of meaning'. Everyday life is the common-sense world, the world of normal reality that we share with other people, and generally think of as most real. Each of the other enclaves or islands has its own specific cognitive style which differs from that of everyday life but is consistent within its own boundaries. The specific cognitive style of each will have its own specific forms of self-experience, specific forms of suspension of doubt, different forms of consciousness or attention, of time perspective, of sociality. Schutz gives as examples of such finite provinces of meaning: the world of dreams, of the theatre, of a child's playing, of a scientist's engagement in theorising, any intense aesthetic or erotic experience (Schutz, 1973; Berger, 1997).

Imagine then 'the state of being fascinated' as a finite province in itself. Perhaps one of its specificities is an enhanced ability to feel the energy of an idea, like a great wave, coming towards us. Another would be the heightened ability to sense two thoughts at the same time and perhaps for a timeless moment to dwell in a wordless gap between them. Fascination is always speechless, though it may be followed by a torrent of words or laughter.

Peter Berger (1997), inquiring into the human experience of laughter, speculates that experiencing something as comic is similarly leaping into another finite province of meaning, where, as in the world of dreams or play, different logics apply: the categories of time and space, the relation to oneself and to others – all are different. He notes (1997, p. 47), 'both laughter and weeping place an individual in marginal or border situations. Man's eccentric position allows man to perceive the world as both constrained and open, as familiar and strange, as meaningful and meaningless. . . . Man is essentially a marginal being.' Where are we more likely to find the need for a transportable environment than in a marginal place?

I put forward these ideas from Schutz and from Berger not as explanations but as tentative descriptions of the state of mind that 'transportability' might bring. Of course, not everything that is transportable is fascinating. A personal memory of an Indonesian family gently creating a place to spend a day on the beach by suspending everything they had brought (the baby, the lunch, the water bottle) from a tree, using lengths of fabric that moments before were part of their dress, is an example that resonates pleasurably with spatiality and simultaneity, ingenuity and thoughtfulness.

A further important idea here is the way in which the many aspects of 'play' that are implicit in the transportable (for example, in the engineer's sense of having space or freedom to move, or the linguist's play of words, or the performative sense of activating or using) are life-enhancing and bound up with our deepest human purposes. The psychoanalyst Donald Winnicott stresses the creative profundity of play:

> Creativity is inherent in playing, and perhaps not to be found elsewhere. A child's play may be to move his head slightly so that in the interplay of the curtain against a line or the wall outside, a line is now one and now two. This can occupy a child (or an adult) for hours. (Winnicott, 1986, p. 64)

Elsewhere Winnicott explains that by creativity he refers not to 'the successful or acclaimed creation' but rather the meaning that refers to a colouring of the whole attitude to external reality. '*It is creative apperception more than anything else that makes an individual feel that life is worth living* [author's italics]' (Winnicott, 1971, p. 65).

Winnicott (1971, p. 39) cites a statement by Marion Milner:

> Moments when the original poet in each of us created the outside world for us, by finding the familiar in the unfamiliar, are perhaps forgotten by most people; or perhaps they are guarded in some secret place of memory because they were too much like visitations of the gods to be mixed with everyday thinking.

The poetic insights of play and of fascination sit outside everyday thinking. We go there, but we cannot stay there. Part of the work of creation entails finding, or recognising, the familiar in the unfamiliar.

Having planted these thoughts – of dualities, shifts and moments of delighted attraction – and tried to link them to ideas of creativity, vitality and creative apperception recognised while moving outside 'everyday life', I must now shift focus to the application of the transportable in a perhaps unexpected context – that of stage magic and illusion and the relationship between some older spatial amusements and today's new spatial disciplines. In fact attributes of transportability were harnessed by this genre of early twentieth-century entertainment in ways that deliberately set out to create a sense of magical engagement and delight for their audience. Transportability was either the hidden secret behind the magical effect or, conversely, the semblance of transportability was deliberately evoked in order to play upon the perceptions of the audience.

The practical applications of transportability within stage magic routines were used to fascinate the audience with the thrill of seeing something appear, or disappear, or turn into something else. These examples relate to audiences and performances from a past time; the acts described are old-fashioned ones and all this material is readily available. However, the routines still have the potential to suggest new thoughts about space to any designer who is concerned with performance and experience rather than style.

Most obviously, the methods used to achieve the productions, disappearances, transformations and transpositions can suggest ingenious ways to move things from here to there. The apparatus was often designed to conceal a transportable and secret load, either out of sight or *beyond our perception* in the right circumstances. This would allow people or objects to be carried on and off the stage in secret (Figure 6.1).

Often the apparatus was able to collapse, fold up or suddenly expand by articulating in unusual ways, or making refined use of the properties of its materials such as elasticity, malleability, compressibility or high tensile strength. It is interesting to consider some of early twentieth-century designer Eileen Gray's furniture pieces in comparison, bearing in mind her interest in flexibility. Caroline Constant (2000) points to Gray's reinvention of eighteenth-century secretaries (the name refers to the secret compartments that characterise these writing desks) as simple compact furnishings with movable parts that must be manipulated to disclose their purpose.

For their performance, many tricks required the use of apparatus or equipment (which could be contained in a surprisingly shallow space) that had to be conceived with an accurate and inventive understanding of the dimensions and capabilities of the human body. Our preconceptions can prevent us from realising that a form that appears to be made up from self-contained and separate elements might possibly conceal a single continuous volume (Figure 6.2). Some apparatuses used topological manipulations to move space from here to there. Consider a simple trick like a 'Fall-Apart Dove Vanish', where a pocket of space flips outside the perceived boundaries of the overall form. The dove or other item that was intended to disappear was placed in a pocket that starts out on the inner face of a receptacle but then ends up on an outside face, which was not seen by the audience. In architectural terms this concept is similar to a volume of space that is designed so that it is perceived as being outside the building envelope, often to fascinating effect.

The magician's clothes were often spatially complex. The stage magician's costume traditionally carried a plethora of secret spaces, poacher's pockets, inner cavities with vertical openings, pockets tailored within the sleeves or in the flaps of the tailcoat. ('The decline of the waistcoat has affected magic more that the invention of the communications satellites', wrote Henry Hay (1982, p. xi) in the introduction to a revised edition of his book.)

New routines often correlate in date with changes in available materials. New ways of achieving magical suspensions and levitations followed the development of processes for producing piano wire of greatly improved tensile strength in the 1860s. Perhaps the same desire for balance that produced all the varieties of cantilever chair in the 1910s and 1920s had already inspired the series of cantilevered devices which permitted some of the other forms of magical suspensions that were so characteristic of the late nineteenth century (Figure 6.3). The history of the bicycle and the history of stage magic also intertwine, with ball bearings, India rubber and tubular steel framing bringing qualities of lightness and effortless movement to the performance of each. Tempered steel springs and elastic cords, collapsible wire skeleton frames or inflatable forms dressed with ultra-light silk fabrics, all contributed qualities of speed and lightness to folding, fanning and pivoting elements and often gave the appearance of 'transparency' in all its ambiguous senses (Figure 6.4).

In general, the old-fashioned apparatus, considered alone, is more fascinating to us today when we know its purpose and therefore what latencies it

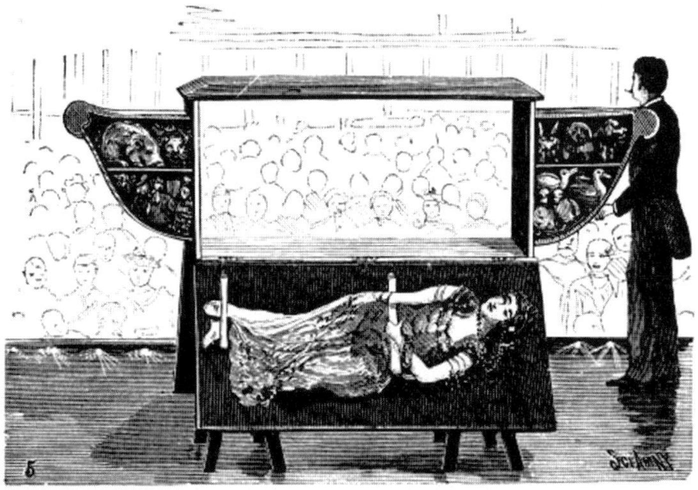

THE MYSTERY EXPLAINED.

TAKING OUT THE BIRDS AND ANIMALS.

6.1 'After the Flood' – quantities of live animals and a mysterious female appear from the 'empty' ark (Hopkins, 1898)

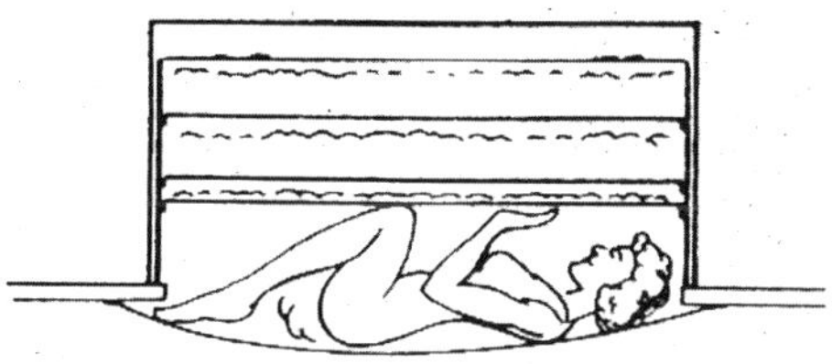

6.2 'The Dressbox Illusion' – successive boxes are removed from the table, leaving a final box that appears too shallow to contain a person (Sharpe, 1992)

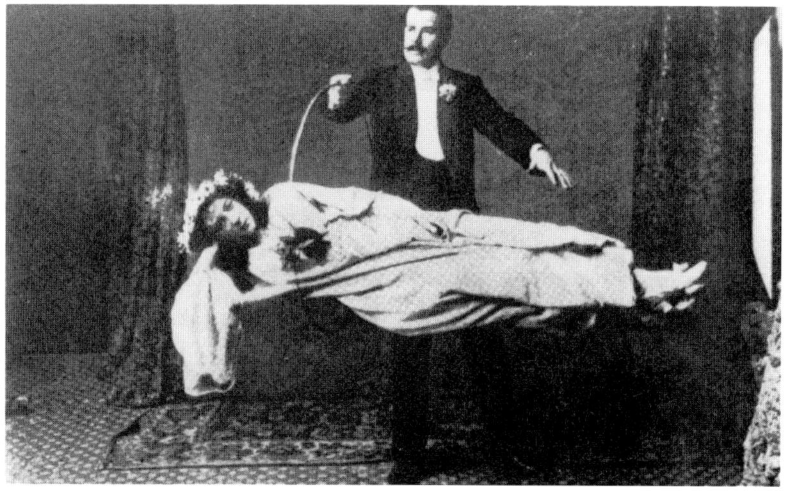

6.3 David Devant performing the 'Sylph' levitation (in which a cantilever plays a part) (Dawes, 1992)

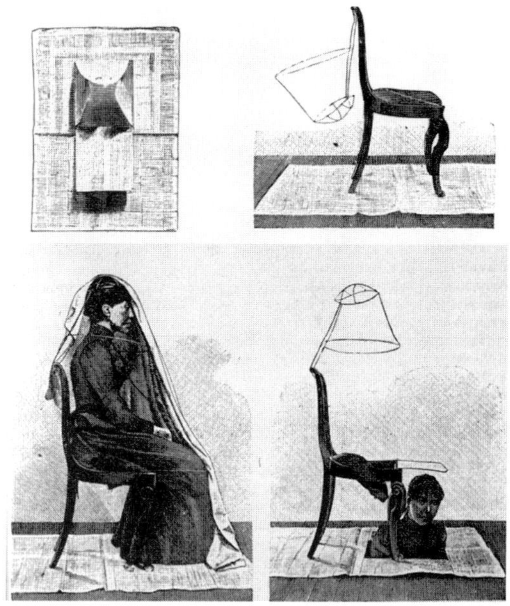

6.4 Buatier de Kolta's 'Disappearing Lady' (Hopkins, 1898)

holds. We would like to play with it ourselves, to press the springs and open the secret drawers. In this it has the charm of any multi-purpose gadget, the potential to be many things and to have multiple functions.

The stage magician may be considered as the combination of himself and his apparatus, something like the concept of a bicycle plus its rider as a single powerful unit. Consider the mobile spaces surrounding the stage magician that are activated through performance as living prostheses. Starting closest to the body, formed by the crevices of flesh and the naturally adhesive traits of the skin, are all the little places that come into existence through the flexing of a limb or the slight cupping of the hand. The next layer lies within the clothes, and in the activation of the pockets and folds that I have described. Next we have artificial spaces very near the body, such as within false fingertips, and all the elasticated retractions known as pulls (Figure 6.5). Then there are the spaces that come and go within the apparatus as it is deployed, opened out, expanded and so forth. A piece seen in one way appears quite different when it is turned on its axis. The meaning or apparent purpose of an object may shift as the performer changes his orientation towards it (Figure 6.6).

Next, the spaces of magician and apparatus combined may be used to carry items on and off the stage, often in linked sets of manoeuvres, as the load is passed from point to point in a chain of moves. Then we have the transportable spaces of the magician's gestures, directing our attention here or there, emphasising emptiness with the passage of a hand or the solidity of something with a rap. A movement of the magician's eyes shifts the balance of our attention. Then we have all the other spaces that surround the performer – the 'interior' sanctuary created by the spotlight, the outer darkness beyond the magic circle of light, the ambiguous and uncertain spaces created in the camouflage of the curtains and backdrops, the spatial misreading of depth, the places that seem to come and go in the shifting veils of darkness (Figure 6.7). Yet further away are the spaces above and below the stage and in the wings, and all the concealed thresholds and devices that activate them, such as the traps, wires and hydraulics of the theatre, the gauzes and scrims and tricks of lighting that allow solid bodies to materialise out of the ether (Figure 6.8).

Beyond these (and outside the scope of this chapter) are the transportable and transporting spaces of the carnival sideshow and the funfair, whose descendants continue to appear in new guises in spatial designs of today, often in the context of the art gallery, museum or public monument.

Conclusion

The renowned stage magician David Devant defined magic as: 'the feeling that we have seen some natural law disturbed' (1909, p. 8). Bearing in mind my earlier discussion of excursions from the common-sense world to other provinces of meaning where the rules are different, this is nicely provocative.

A less self-conscious discipline than architectural theory is that of conjuring theory. Henry Hay (1950, pp. 2–5) wrote in *The Amateur Magician's Handbook*:

> The central secret of conjuring (and of art and literature and politics and economics) is a manipulation of interest. . . . What in turn is interest? Interest is a sense of being involved in some process, actual or potential. Interest is not the same as attention. . . . Interest is selective, an expenditure of energy by the interested party. . . . Interest is projected by the interested person, not by the 'interesting' object. Next step for the performer to grasp: *Perception too originates with the perceiver, not with the object*.

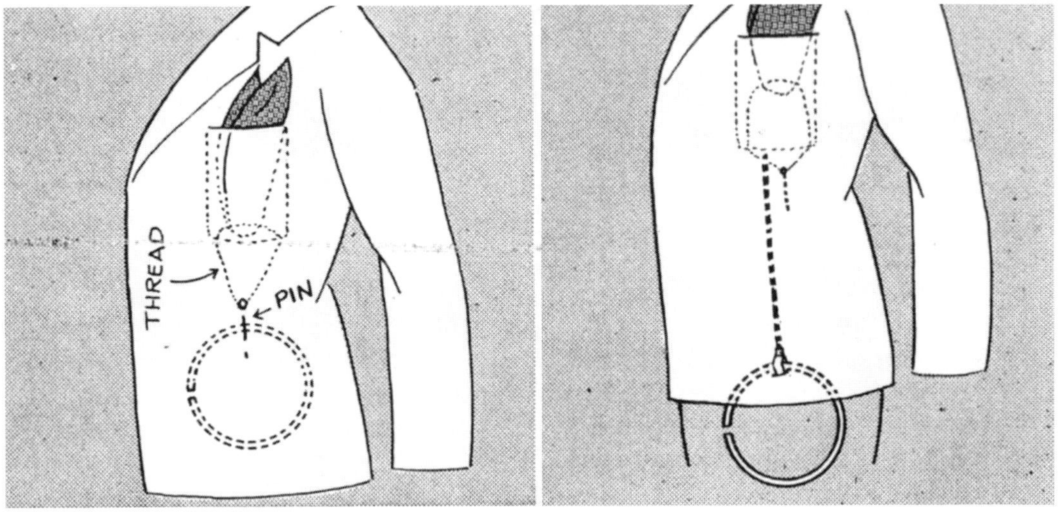

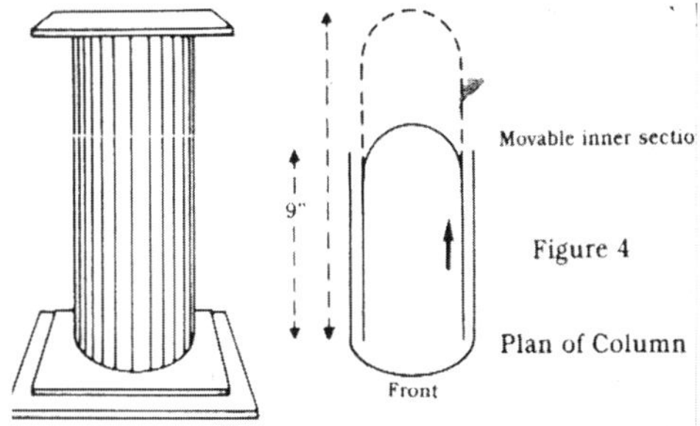

6.6 A deceptively narrow column that may increase in depth without the audience noticing (Sharpe, 1985)

6.5 Secret movable items concealed in a magician's garments (Baker, 1946)

6.7 'A wave of the wand and a table appears' during a Black Art routine (Hopkins, 1898)

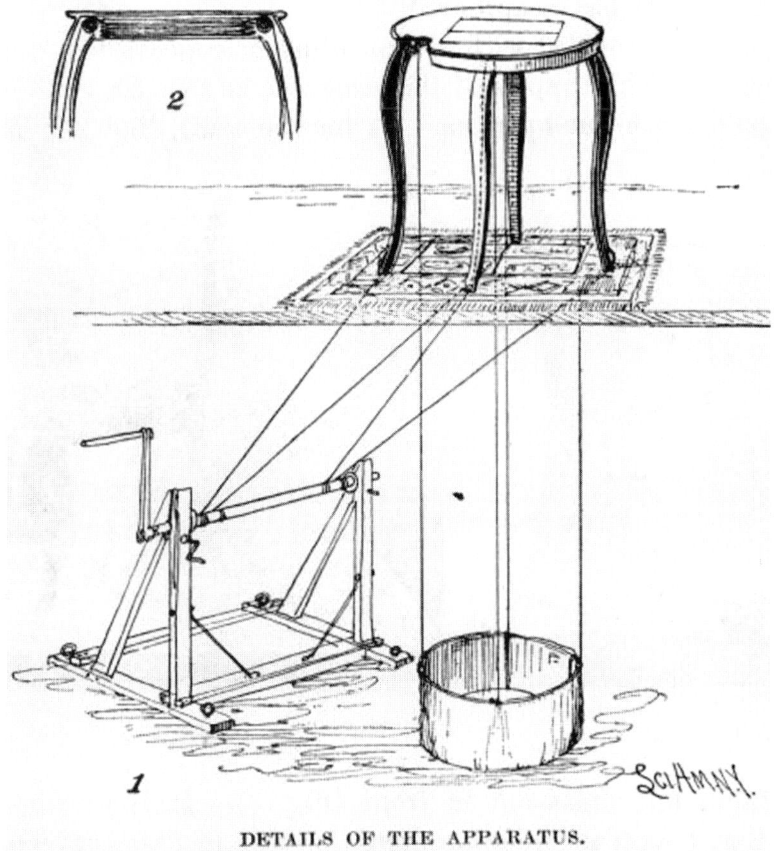

6.8 Below-stage secrets of 'The Appearing Lady', another routine originated by Buatier de Kolta (Hopkins, 1898)

So it is in the audience's perception that the act sits. Such feelings and perceptions have become bound up in our contemporary sense of spatiality, and have returned to the foreground in today's spatial disciplines, with their renewed interest in spatial perception and experience rather than in style, the renewed focus on the kinaesthetic, tactile and sensual, and the expansion of the scope of spatial design to include within its realm the design of an event, or a moment of time, or a visceral sensation. A most noticeable shift in emphasis is the move away from the harnessing of these sensations to produce laughter or screams (as with the tricks of stage magic and other spatial amusements) to incorporating them in designs that are directed towards a sense of beauty or profundity.

This chapter draws together Schutz's proposition that our mental life is made of a series of excursions out of the common-sense world and the natural attitude into other self-contained provinces of meaning and back again, and the belief (following Winnicott and Berger) that these excursions are creative and ultimately life-enhancing. We can recognise that the transportable is literally extra-ordinary (outside ordinary life) and that our recognition of this is itself the opportunity for creative apperception and the sense that life is worth living. Our pleasure in 'transportability' is bound up in our joy in language, and our ability to create understandings. If the sensation of magic is 'the feeling that we have seen some natural law disturbed' then brief excursions that carry us away from natural life have the potential to offer us the pleasure of magical experience. In one modest genre of popular entertainment, the mechanisms of transportability have been used to offer us this pleasure and in this context the 'transportable' often makes us laugh.

References

Baker, A. (1946) *Magical Ways and Means*. Minneapolis: C. W. Jones.

Berger, P. (1997) *Redeeming Laughter: The Comic Dimension of Human Experience*. New York: Walter de Gruyter.

Constant, C. (2000) *Eileen Gray*. London: Phaidon.

Dawes, E. (1979) *The Great Illusionists*. Newton Abbott: David and Charles.

Devant, D. (1909) *Magic Made Easy*. London: Cassell.

Hay, H. [1950] (1982) *The Amateur Magician's Handbook*. New York: Harper & Row.

Hopkins, A. (1898) *Magic, Stage Illusions, Special Effects and Trick Photography*. New York: Munn and Co.

Rowe, C. and Slutzky, R. [1964] (1997) 'Transparency: Literal and Phenomenal', in *Transparency*. Basel: Birkhäuser Verlag.

Schutz, A. (1973) *Collected Papers 1: The Problem of Social Reality*. The Hague: M. Nijhoff.

Sharpe, S. (1985) *Conjurors' Optical Secrets*. Calgary: M. Hades International.

Sharpe, S. (1992) *Conjurors' Mechanical Secrets*. London: Tannen's Magic Manuscripts.

Wilk, C. (1981) *Marcel Breuer: Furniture and Interiors*. London: Architectural Press.

Winnicott, D. W. (1971) *Playing and Reality*. London: Tavistock Publications.

Winnicott, D. W. (1986) *Home Is Where We Start From: Essays by a Psychoanalyst*. New York, London: Norton.

Ephemeral Landscape, Portable Dwelling: The Ice Fishing House and the Fish House Community

Martha Abbott
University of Minnesota

A seasonal nomadic phenomenon occurs on Minnesota lakes each year, and that is the brief appearance and evanescence of ice fishing communities. During the three coldest winter months every year an ephemeral frozen landscape gives rise to innumerable temporary communities of ice fishing houses and their inhabitants. Like a circus coming to town – hoisting huge tents, convening collapsible cages, and establishing temporary territory – these portable fishing villages conjure up an almost magical allure brought about by such wholesale transformation. Individual ice houses are designed and built as transportable structures. Fishermen equipped with skills evolved over many childhood winters build and place their ice houses without the help of designers, urban planners, or architects. Each winter when lakes freeze to a suitable thickness, these nomadic men and women transport their fishing houses onto the ice. Sites are chosen, houses skidded into place, and snow pushed around perimeters and watered to freeze into both footing and purchase. As warm weather approaches and the ice darkens, houses are broken loose from their frozen moorings and trailered off the ice until the following year. Because of the ephemeral nature of the landscape, the houses are necessarily temporary; they cannot persist year round. By springtime, no discernible evidence of these winter communities remains.

Premise and intentions

This chapter explores the notions of portability, transience, and permanence by looking at this intriguing contemporary phenomenon and its relevance for deeper social meaning. This seasonal practice assumes significance as an occasion to challenge perceptions of what is possible or desirable by piquing our interest in the role of portability within design disciplines, and by reinvigorating the thinking that informs the production of architecture and, by extension, community. A closer look at this vernacular activity offers an opportunity to look beyond the discourse of esthetics and form, to deeper issues within the structure of architecture and urban considerations. By probing beneath the surface of the physical and formal characteristics of these icy villages we may seek opportunities for redefined social and ethical possibilities that emerge from this alternative social order.

One could imagine these wintry villages not as prototypes or models to imitate, but as means to probe, question, and examine our grasp of the issues of transportability, placelessness, permanence, and ephemerality that may offer an unintentional perspective on a critical view of eternal human questions. Each winter these nomadic fishermen and women brave subzero temperatures to live for days at a time on frozen landscapes, only to be eventually forced from their short-lived moorings to more conventional dwellings when warm weather sets in. The question is – why? What is the allure of this Spartan existence? Though ice fishing is a recreational activity, men and women choose to leave conventional houses and spend long weekends living in cramped, cold, dimly lit fish houses. Why? Can these funky settlements serve to open up possibilities for speculation about how we live and why? It would be too easy to make an esthetic

7.1 Ice fishing in Homer, Minnesota

7.2 Night fishing

phenomenon of these portable villages; on the contrary, let us consider why these settlements have such appeal. What do they tell us about our social and, perhaps, political intentions? It can be suggested that these communities are instances of transgression. Unintentionally, these nomads transgress the contemporary inability to surrender to environment, context, and circumstances under which architecture is materialized. Their transgression permits us to reflect on the uncertainty and mutability of our social needs and desires. Is the ephemeral nature of the landscape important to its vitality and consequent appeal? Are we able to embrace and reveal the transient nature of this place and its essential character? Thoughtful observation of this phenomenon could enlighten our pursuit of understanding the roles of transience and permanence in our current social structure. This vernacular building practice and *ad hoc* emergence of community may hold lessons for designers today, not in the forms adopted and constructed or in the methods employed, but in the unintentional questions inspired and the vision revealed.

The disposition and configuration of these transitory villages has nothing to do with good views, interesting terrain, or proximity to amenities, but, ostensibly, with what is happening beneath the surface: good fishing. Uncles, mothers, brothers, and cousins cluster their houses in pockets, forming neighborhoods; one-lane roads wind between buildings and stretch from cluster to cluster across frozen lakes; acres of unclaimed ice land allow settlement anywhere. No one owns this ice land (because it is not technically land); one possesses it by being there. One needs only a fishing license, a fish house, and a trailer hitch. Houses are not technologically advanced, size is limited to what can be drawn behind a vehicle, and walls are often not insulated. Generally stocked with a stove, card table, a few chairs, and some bunks, a house typically sports one or two fishing holes, sometimes four. Constructed largely through a vernacular *ad hoc* process, ice fishing houses are provisional structures, often shanty-like in their construction. However, these details are incidental to the assertion that if we content ourselves to look merely at the formal and material properties of the ice houses or even the collective configuration of houses, the critical import of these settlements may be overlooked, leaving the underlying significance of their persistence unaddressed.

Communal experience in an ephemeral landscape

The settlements are significant as experiences of community, social interaction, and shared experiences. This runs counter to the individual private habitat of conventional, titled dwellings to provide communal social experiences and a social dimension to dwelling. Partaking also means simplifying and stripping one's life down to the basic, favored necessities. These portable villages offer a different sense of personal space and communal space. The 18 by 30 meter (60 by 100 foot) plot of land on which to build the typical single family dwelling according to building zoning is moot; ice houses are placed as close to, or as far from, the "neighbor" as feels socially and sportingly appropriate. Fishermen and women usually locate themselves near friends or relatives and always in good fishing spots, generally near deeper parts of the lake.

Considering ice houses as provisional structures offers a more positive spin on their temporal nature, in that their provision for a particular period of time enhances and intensifies the ephemeral circumstances under which they come into being. Unlike cabins, for instance, which are featured in a proliferation of publications touting the charming, the quaint, the rustic, and the palatial, ice houses are rarely featured, with the exception of the odd local newspaper highlight (for example, see the article on ice fishing in *Star Tribune*, Minneapolis, February 2004). Since we tend to privilege the

7.3 Ice fishing settlement

7.4 Pitching a plastic tent on the ice (two images)

permanent and the owned, these potential instances of social significance are often overlooked. Primitive and *ad hoc* as these ice settlements are, without doubt they throw light on social and political questions of ownership, permanence, and social order. In a culture of mindless consumption, obsessed with ownership, property, and permanence, the collective denial of our temporal human condition has the effect of diminishing our social awareness.

Communing with nature: magic, memory, and surrendering

The magic of this urban vernacular lies in the ritual of returning to a quiet place that exists for only part of the year. It lies in the comfort of hibernating within familiar walls. It lies in the simplicity of just being there. Out on the frozen lake one occupies a thin stratum with an immense dome of sky above and deep waters below. Beneath the ice are unknowable depths, and the only connection between the human realm and the nether realm is a small hole in the ice. Simultaneously frightening and exhilarating, one is suspended between these stratified worlds. The cold conditions are a compelling aspect of the allure of these communities. The ice fishing house is not about sensual pleasure, but is, perhaps, about the instinctual and unselfconscious connectedness to nature that these nomads inadvertently find missing in their conventional worlds. These ice dwellers escape to a simpler and sparer plane of existence – an edge condition – where they can feel something they are not permitted to feel in their insular, permanent lives.

Each year these ice fishing villages appear and disappear, never the same and yet always the same. As the winter fishing season closes, houses are dismantled, collapsed, deflated, bundled, broken down, folded, or disassembled and stored in garages, along alleys, and under tarps until the following year. The memory of these villages lingers through the warm spring, summer, and fall months, awaiting the frozen landscape. The fulfillment of a desire to engage in this way of life occurs during a brief interlude each year; memories of past interludes and anticipation of those that will be, fuel desire for engaging once again in this wintry pastime. The ice is laid bare beneath vast, crisp skies, and so are its brief inhabitants. Time spent in this barren environment allows a temporal freedom from the responsibilities and underpinnings of conventional life. In the act of simplifying, we apprehend the ephemeral nature of our own existence; we are, in effect, stripping away conventional insulation against our own mortality.

Not an esthetic phenomenon

Considered formally, ice-fishing houses are not especially charming or seductive. At best they are quirky and some are even downright ugly. In our contemporary architecture culture, the "funky vernacular" notwithstanding, there is a preoccupation with how buildings look. Images of architecture are particularly susceptible to an esthetic rendering. In his book *The Anaesthetics of Architecture*, Neil Leach critiques our preoccupation with images and image making and suggests that:

> While aestheticization remains a background cultural condition that permeates – to a greater or lesser extent – the whole of present society, its effects will be all the more marked within a discipline [like architecture] that operates through the medium of the image. . . . This privileging of the image has led to an impoverished understanding of the built environment, turning social space into a fetishized abstraction.
> (Leach, 1999, pp. 9–10)

I make no attempt here to construct a vernacular esthetic of these primitive, rough-hewn fishing houses, though it could easily be done. Instead, I think this investigation is an opportunity to see beyond the discourse of esthetics to deeper architectural issues. Probing beneath the surface of the physical dimensions of these ice houses offers an opportunity to redefine our social and ethical awareness. In fact, as Leach suggests, though

> [g]ood design obviously depends upon a strong sense of visual awareness, . . . this emphasis on the image has certain negative consequences; and it is in a discipline like architecture, which is so directly involved with social concerns, that these negative consequences are likely to be most keenly experienced. The aestheticization of the world induces a form of numbness. . . . What is at risk in this process of aestheticization is that political and social content may be subsumed, absorbed, and denied. The seduction of the image works against any underlying sense of social commitment.
> (1999, p. 45)

Estheticization then distances us from social and political realities, undermining my intent. In fact, what seems so clearly evident here is that there is a political and social verity manifest *in the very existence of the communities themselves*, reappearing each year without titles or taxes.

Freedom and lightness

The freedom from what holds us responsible and a willingness to escape our rootedness in the conventional denial of the temporal human condition put us in touch with the fundamental aspect of our existence: that we are mortal beings. As Alberto Peréz-Gómez suggests, ephemeral architecture should allow us to think the eternal and yet remain light and unencumbered. He implores us to remain conscious of our sensuous bond to the earth, recognizing and coming to terms with the human condition of death (an idea discussed by Professor Alberto Pérez-Gómez in his keynote lecture at the Third Transportable Environments Conference, Toronto, Canada, April 2004). The ultimate ephemerality is our own mortality, something we attempt to conceal. The ephemerality of the ice season, and the communities that subsequently materialize on this landscape, though not consciously motivated, offer the sensuous connectedness between self and environment that Péréz-Gómez describes. These present-day nomads certainly make the best of the climate, but also, untethered from obligations to the conventional world, these hearty souls feel strangely "at home," unwittingly embracing their own impermanence.

Curiously, the ephemeral landscape on which these communities gather is territory unowned, untitled, and unpossessed. Could a compelling draw to this harsh landscape be the *lack of* ownership, taxation, or economic base on which to circumscribe rules and regulations about its use? In a culture whose myths regard as sacred the ownership and possession of land, ideas, technology, and material goods, it is extraordinary to encounter a situation where "land" disappears and reappears seasonally and, hence, cannot be consigned to the cultural forces that shape our current condition of eroded social function. This landscape does not conform to the customary means and methods of territorial ownership; it is a veritable *tabula rasa*. There is a primal human desire to engage creatively with the world that these settlements ultimately represent. These encampments might serve as a conceptual window into the nature of human desire for an alternate social order arising from their essential connectedness to this place. Again, these ephemeral villages facilitate the possibility of forging a link between the self and other, and between the self and the environment. It is, thus, strange and hopeful to find, within our culture of ownership, this unexpected yet thriving island of communal possession and dispossession.

7.5 Paneled interior with two holes

7.6 Two huts

Permanence, ephemerality, and the temporal human condition

Our sense that mobility and transportability are recent phenomena only appears revolutionary, Pérez-Gómez reminds us, because we come out of a culture that insists on permanence and stability, and, he adds, traditional architectural theories, either narrative or constructed, were always mindful of the ephemeral, temporal condition on the earth. Ephemeral structures reveal a poetic possibility that intrigues us. We, too, are here only temporarily. Do our contemporary architectural theories and institutions acknowledge this? Western culture, obsessed with permanence, private property, and the production of durable, lasting architecture unfortunately conceals the ephemeral nature of the human condition. Our institutions are predicated on a stable and predictable ground on which to enact our daily affairs. Because architecture embodies our social, cultural, and political beliefs and values, when these beliefs become oppressive and burdensome, our architecture reflects this. For architecture to inspire, according to Pérez-Gómez, however, it must also remain the repository of memories, hopes, and myths, keeping the collective imagination alive.

Ephemeral landscapes and portable architecture allow us to consider the eternal questions and yet embrace lightness. What is this lightness in the context of the ice fishing environment? Lightness could be described as a surrendering to conditions and circumstances within which we find ourselves that reveals something about both our environment and our selves. This means that we accept and embrace the transient nature of our selves, our environment, our existence, and our mortality. It is an ethical choice whether or not to inhabit this territory. We must accept the limits of the environment and the limits of the human condition, if we are to embrace the lightness of change, instability, and placelessness and come to terms with the human condition of death. Perhaps these quirky shanties, patched together by a few hearty adventurers, serve to bond us to the earth and to the natural human condition.

Technology and creativity within limits

Technology today is obsessed with overcoming constraints. However, real constraints are the seedbed for creativity, as constraints open up possibilities for the human imagination to flourish. Architecture, too, happens within constraints and limits – an essential part of the human condition. In deciding whether to inhabit and how to engage the landscape and construct shelter, we may or may not use a highly technological solution; that is a choice. There is a danger, however, in the seductiveness of pure technological problem solving. If we step back, we find that the juxtaposition of technological seductiveness seeking to obliterate constraints on the one hand, and consideration of fundamental questions on the other, becomes polarized around political, economic, and social structures that stand to benefit from our waning attention to the essential questions. Can we simultaneously keep these questions in the forefront and maintain the focused energy to develop technologies that allow us to solve problems that we determine require transportable environments? These fishermen have determined a way, with limited technological solutions, to inhabit the ice and (though temporarily) fully immerse themselves in the profound, quiet connectedness of an otherwise adverse environment.

Desire and the poetics of revealing place

If we are, indeed, committed to engaging our environment, and whatever conception of community we propose is only temporary until the earth turns, the season changes, the ice melts, the ground dissolves, and all must be disassembled, collapsed, moved, erased, or allowed to become something else, there remains an opportunity to uncover the inherent character of the place and our connectedness to it. Portability is seen here not merely as a capacity to be easily moved, but as an architectural possibility that can reveal and celebrate the transitory nature of its context. In this instance, the notion of portability is dependent on and engaged with the changing seasons. Occupying this frozen ground is a way of inscribing the territory to reveal a poetic vision and allow the human condition to comfortably occupy a place where imagination thrives. Portability is important precisely because it acknowledges the temporal context, revealing that context and the circumstances surrounding the materialization of our architectural constructions.

There is an ethical dimension to the choice to locate a community in an ephemeral landscape; that is, we must fully appreciate and acknowledge its temporal nature. In this northern climate lakes freeze for months at a time, creating an ephemeral *terra firma* that can be inhabited. The ground is seductive because it is impermanent. The poetic potential to reveal its essential seasonal character by briefly and lightly inhabiting the ice is profound. As human beings we can explore, reappraise, and momentarily reveal the inherent properties of this place: its climate, its nature, its poetic potential, and, ultimately, our connectedness to it. Architecture opens up the space of desire, as Pérez-Gómez suggests, and this is critical for the human condition. As the ice fishing phenomenon reflects, the possibility for communal dwelling, even temporarily, manifests our human desire to engage with our world.

7.7 Pick-up truck on thawed ice road

Reference

Leach, Neil (1999) *The Anaesthetics of Architecture*, London: MIT Press.

Design

Transformation in Architecture and Design

Chuck Hoberman
Hoberman Designs Inc.

I believe that *transformation* is a major new frontier of design. For me, this word refers to an object or structure that *transforms itself by itself* – one that has an innate property of controlled change. A transforming object may be foldable, retractable or shape shifting. Such capabilities lead to functional benefits: portability, instantaneous opening and intelligent responsiveness within the built environment. This chapter explains this theory of transformation and the principles that form the foundation for this new field. These principles touch on transformation in nature, how the behaviour of objects emerges from their underlying geometry, and how that emergent behaviour leads to an astonishing variety of design possibilities from the microscopic to the architectural scale.

Transformation theory

When we apply pressure to an object, it may respond by bending, breaking, squashing or resisting inertly; however, many other responses are also possible. Specific behaviours can be designed into an object, behaviours such as expansion, dilation, controlled folding, shape change – in short, *transformation.* The creation of transforming objects requires a new design theory, a conceptual framework that draws on mathematics, mechanics and structural engineering. It is an approach that is inspired by nature, not from a visual standpoint, but rather from a functional one.

How does this approach differ from existing design practice? After all, many objects are designed to transform their size and shape. There are numerous practical reasons to make objects transform, for instance portability, compacting packaging for shipping, and to make possible multi-use products. Some examples are camping tents, balloons and umbrellas. Each of these products has its particular way of changing configuration, whether through assembly, inflation or unfolding. While the designs shown are functional and even elegant, their transformation capability is a kind of 'add-on' to their basic design. My goal has been to develop design methods that are comprehensive and generalized rather than specially created case by case.

The process of transformation

A transforming object has, by my definition, certain unique characteristics. These qualities relate not to its appearance or even its material properties, but to its *behaviour*, a process that is:

- complete and fully three-dimensional
- smooth and continuous
- reversible and repeatable

After years of exploration and experimentation I have found that these qualities are the critical design criteria. They result in specific functional benefits for products such as ease of use, fluid responsiveness and adaptability. Such criteria lead to an integrated design approach that provides the capability to build transforming structures at large scale where structure and mechanism are combined.

It is actually quite rare to find instances of this complete, smooth, continuous transformation in designed objects. For example, a tent is erected in a series of discrete steps. An umbrella changes its form in a continuous manner; however, its change is not complete in that its overall length is not altered. A balloon may be inflated smoothly and completely, but when deflated it contracts in an uncontrolled manner. Smooth and continuous transformation, while rare in man-made objects, is common in the natural world.

When we watch the changing shape of clouds on a clear day, the curling flow of a turbulent fluid or a time-lapse film of a growing plant, we are observing natural transformation: forms found in nature that metamorphose into new forms. These transformations in size and shape reveal subtle and ordered natural processes. Natural transformation embodies these key qualities: it occurs continuously and is not broken into discrete steps; it is fully three-dimensional; it moves between completely different sizes and shapes. Reversibility is, however, not found in nature since natural transformation never repeats and is not reversible.

The basis of natural transformation is found at smaller scales. Science teaches us that matter is made up of small units – cells, molecules, atoms and so forth. As we go down in scale, a point is reached where these units are, in essence, indivisible. It follows that what appears as continuous transformation must be comprised of movement between indivisible units. For example, biological growth results from molecular movement and recombination.

This insight – that transformation is the aggregate of many incremental movements – is an essential factor in the design of transforming objects.

Design principles

My work has centred on a fundamental idea: that a designed object can transform in the way a natural organism does. The design of such transforming objects is based on several key principles – a transforming object:

- is made of many parts that act as an integral whole
- has a dual nature – half structure, half mechanism
- is made up of linkages having unique properties – I call these kinetic blocks
- has behaviour based on its underlying geometry

Structure and mechanism

When we apply a force to an object, the response to that force in the object will fall somewhere along a continuum. At one extreme the object resists inertly and eventually the force will result in motion. That motion may start as an elastic deformation. Beyond that it may lead either to uncontrolled deflections (structural failure) or to controlled motion.

By definition, a structure is an assembly of materials intended to sustain loads whereas a mechanism is an assembly intended to convert force into controlled movement. A transforming object is a hybrid between a structure and a mechanism. It is defined as a mechanism because it converts applied force into movement. However, normally mechanisms are not seen as integral objects. Rather, when designing mechanisms or machines, the emphasis is on producing trajectories to achieve a particular function.

While a transforming object is, in part, a machine, it is more than an assembly of individual links – it is also a complete unit that sustains its integrity and form. In this sense it is a structure. Thus it is both structure and mechanism at the same time: the links of the mechanism (transferring motion) are identical with the structural elements (providing support and shelter). It can be solid at times, flexible at others.

Kinetic building blocks

A useful conceptual tool with which to understand how to design transforming objects is the idea of a 'kinetic building block'. A kinetic building block is, in essence, a linkage that can connect to other linkages. Like its static counterpart, it must transfer forces. However, in addition it must transfer motion to neighbouring blocks. Figure 8.1 shows examples of such blocks or linkages. The edge of the block, shown as a dashed line, transforms in size. This changing perimeter represents the edge at which motion is transferred from block to block.

Transforming objects may be built by assembling kinetic blocks together. Because each block is a linkage, when it is manipulated it transfers force and motion to neighbouring blocks. When these forces and motion are transferred completely throughout the network, there is synchronous movement. Many of my inventions have focused on the discovery of particular kinetic blocks. The kinetic block acts as a kind of DNA for the transforming object, giving it its own unique properties.

Underlying geometry

One of the biggest challenges in designing transforming objects is how to create a wide variety of shapes. When I first became interested in the problem of transformation in the late 1980s, I found that there was no general method to make closed shapes – that is spheres and polyhedra – nor were there solutions for doubly curved surfaces. I felt that without a general method to make different shapes, the possibilities of developing useful designs would be very limited. Thus I focused on this problem and in 1987 I discovered and patented a general technique for making expanding structures of virtually any shape (U.S. Patents 4,780,344 and 5,024,031).

My discovery was, in essence, an algorithm to translate particular surface properties into the angles and proportions of particular linkages. When connected together, such linkages form a matrix that lies on the defined surface. As this mechanical matrix is manipulated it expands and contracts, but the links *continue* to lie on that surface. They grow and contract, but do not change their overall shape. The importance of this discovery is that it can be generally applied and consequently provides a method for making expanding structures of virtually any shape (Figure 8.2).

When an object expands, the points on its surface move radially outwards – this is the relationship between shape and trajectory, a relationship whereby hundreds or thousands of links are synchronized by the geometric organization of the system.

Scale, stability and movement

Perhaps the clearest demonstration of the power of these design principles is in large-scale transforming structures. In my practice I have built a series of such large structures since 1990.

In an analysis of any large structure, the primary issue is stability. Even with conventional structures, stability does not mean rigidity. There is always some movement, but in order to stay within the elastic limit of the members, deflections are small relative to the overall span. However, the movement of a transforming structure is of similar magnitude to its span. Thus the idea of stability needs redefinition – the criterion is not in minimizing deflections, but rather in maintaining operational control. *Stability is a process, not a state.*

Analysing and predicting the behaviour of large transforming structures is, however, a daunting prospect. Tons of material must move synchronously through space, and it is necessary to predict the forces that flow through a linkage made of hundreds or thousands of parts. Despite these

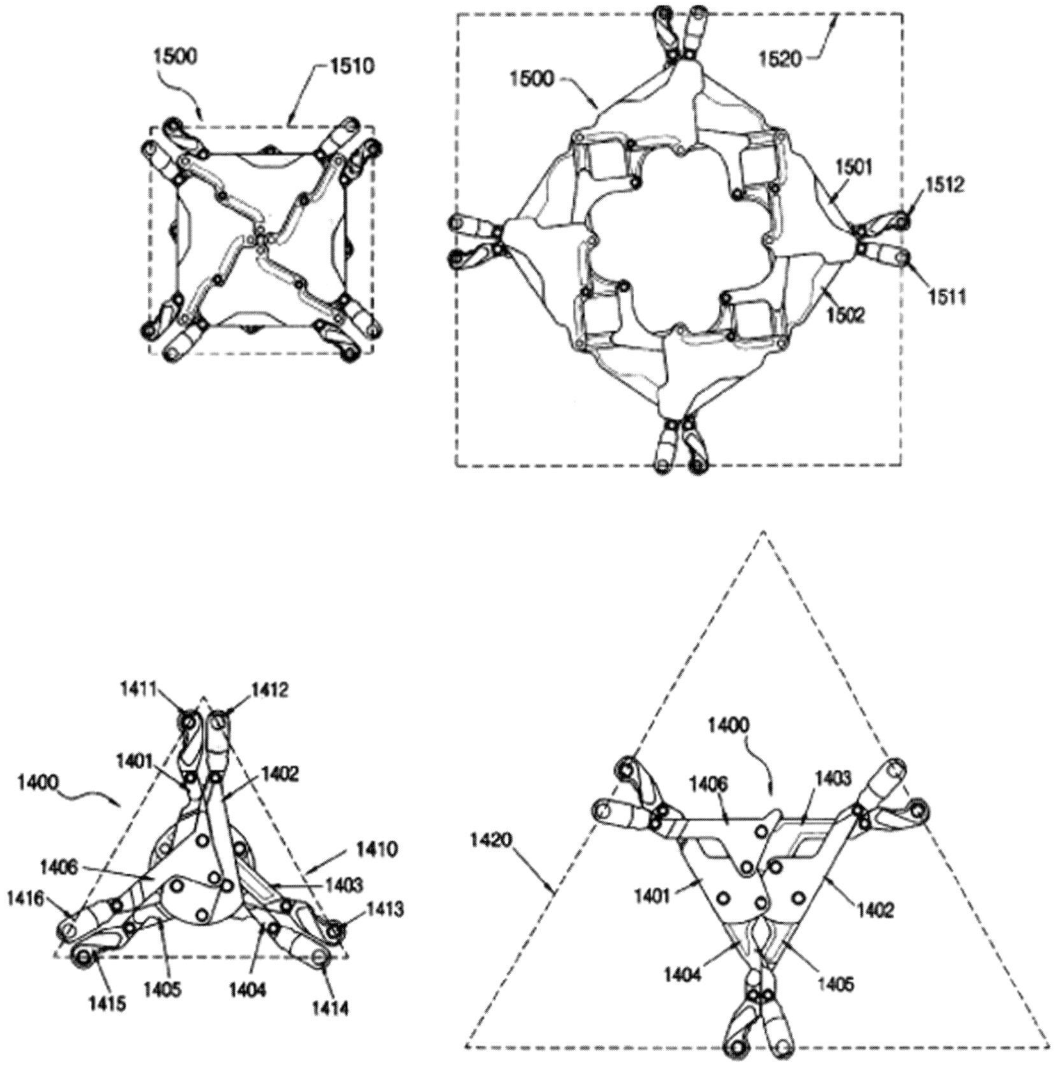

8.1 Kinetic building blocks

8.2 Underlying geometry configurations – cube, tetrahedron, sphere

8.3 Expanding geodesic dome

challenges, in my experience it has proved to be a relatively straightforward task to model the behaviour of transforming structures because they can be analysed for a series of positions, each treated as a 'snapshot'. Standard structural software programs such as SAP or Nastran can therefore provide good results.

Case studies

Expanding geodesic dome

Based on my discoveries, in 1991 I built my first expanding geodesic dome (Figure 8.3). It was deployed by being pulled outwards at the base; as it expanded it kept its hemispherical shape and remained stable. This design can be compared with a static geodesic dome such as those designed by Buckminster Fuller from the early 1950s on. Visually, the overall shape and the triangulated pattern of structural members are the same, but there is a significant difference. Wherever there is a compressive strut in the static dome instead there is a linkage in the expanding dome. Once the expanding dome is locked in position at the base, it 'freezes'. In this state, it behaves essentially as a static geodesic dome, having a similar force distribution.

In 1997, I was invited to participate in an exhibit of architectural engineering called *L'art de l'ingénieur* at the Centre Georges Pompidou in Paris and for this I created a second version of the expanding dome (Figure 8.4). The exhibit consisted of a scale model of an expanding dome with a fabric membrane covering. In its contracted state, the fabric was housed in a cavity at the centre of the dome and when expanded, the fabric became smooth and tense. One of the problems was to maintain consistent control over the fabric as it folds and unfolds. I therefore added extensions onto the links to attach the fabric and these provided sufficient control points so the fabric could be maintained in a separate zone from the linkage at all times.

Expanding hypar

In 1995, I was invited by the California Science Center in Los Angeles to create a sculptural icon for the entrance space to this new museum, then under construction. I proposed an expanding structure in the shape of a hyperbolic parabaloid or hypar for short. This shape is an example of a so-called 'saddle' surface, one that is spontaneously formed by soap films or tensioned fabric (Figure 8.5).

The resulting design was my largest sculptural piece up to that time, an undulating expanding lattice that spanned 15.24 metres (50 feet) when open, folding down to a packaged size of 4.57 metres (15 feet). Although the hypar is a doubly curved surface, it has straight lines that run along certain axes. I exploited this feature by running six tracks along these axes to support the sculpture.

The hypar is made of over 2,000 links and weighs around 2,268 kilograms (5,000 pounds). It is actuated by six cables running along its support tracks all controlled by a single winch. The sculpture is further supported by a single stationary point at its centre. The design has a unique quality in that its form alters when seen from different positions, sometimes taking on a star-like profile, other times appearing in the shape of a bow-tie.

Iris Retractable Dome for Expo 2000

In 1998, I was contacted by my associate, Werner Lorke from the Frankfurt-based company *i/O*, to collaborate on a project for Expo 2000 in Hanover, Germany. Lorke was developing an exhibit celebrating the reconstruction of the legendary Frauenkirche Cathedral in Dresden, which was destroyed during World War II. He wanted to build a retractable dome that would symbolize the destruction and rebuilding of the cathedral. I proposed utilizing one of my inventions, the Iris Dome – so-called because the structure transforms like the iris of an

8.4 Fabric skinned expanding geodesic dome at the Georges Pompidou Centre

8.5 Expanding Hypar

eye. This was my first opportunity to build such a structure at architectural scale (Figure 8.6).

Constructed of high-quality aluminium alloy, the dome's motion resembles a three-dimensional aperture that retracts towards its perimeter supports powered by four computer-controlled hydraulic pistons. The members are connected to hub elements each of which houses four stainless steel pins with paired angular roller bearings. The dome was installed alongside the German Pavilion at the Expo. It sat on 4.57-metre-high (15-foot) columns creating a pavilion space where visitors could gather. Below the dome sat a detailed scale model of the original cathedral. For the six months of the Expo the dome extended and retracted continuously, cycling thousands of times (Figure 8.7).

Looking upwards from beneath the dome, the visitor could see interlocking spirals irising outwards as the dome opened. The sight of the dome forming itself overhead gave a unique spatial experience, relating to our perception of the sky as a kind of dome itself. This radial opening was choreographed with a sound installation describing the cathedral's reconstruction (Figure 8.8).

Mechanical Curtain for Winter Olympic Games, 2002

In January 2001, I was contacted by the Organizing Committee of the Winter Olympic Games to be held in Salt Lake City in 2002. They were creating a venue where each evening the medals would be awarded for all the events of the previous day. I began a creative collaboration with the stage designer, Steve Bass, who wanted a 'curtain' for the stage that would be unlike anything ever seen before. This mechanical structure would have to cover a 22-metre-wide (72-foot) stage and then retract to a compact size to completely reveal the stage. Additionally the structure's movement had to hold the attention of the viewer through its speed of deployment and visual impact (Figure 8.9).

The solution we offered was an iris structure: a semi-circular wall, 11 metres (36 feet) high and 22 metres (72 feet) in diameter, that retracted to form a compact 1.8-metre-thick (6-foot) ring that framed the stage and exposed the medal rostrum. The curtain comprised two main parts: a matrix of movable panels and a static arch that supported those panels. The concept for the screen involved the use of ninety-six panels, each with a skeletal frame constructed from aluminium box sections that were clad in translucent fibre-reinforced panels (Figure 8.10). Four different-shaped panels were used that were radially arranged and layered over each other to form an almost solid screen when closed. The outermost panels attached to thirteen radial slides on the static arch. The lower sections of the panels were supported by 'trolleys' that rode on tracks housed in the stage's turntable located beneath the arch. All movable elements on the perimeter of the arch were directly attached to static supports. The total weight of the moving elements of the screen was about 6,800 kg and the static portion an additional 6,800 kg.

Scenic Technologies, one of the premier builders of mechanical theatre sets and attractions, fabricated the curtain. A prototype section was manufactured first and tested for structural and mechanical performance at their facility in Cornwall, New York. After testing, the completed structure was split into six sections, each fitting on a flatbed, and shipped to Utah. The Mechanical Curtain was reassembled lying flat on the stage, then raised to a vertical position by a series of winches (Figure 8.11).

I commissioned the New York office of Buro Happold to develop a structural system that could resist forces caused by wind and structural weight. The first stage of the analysis was to set a design criterion for the structure that offered the least risk,

8.6 The Frauenkirche Cathedral and the Iris Retractable Dome at Expo 2000

8.7 Iris Retractable Dome piston mechanism

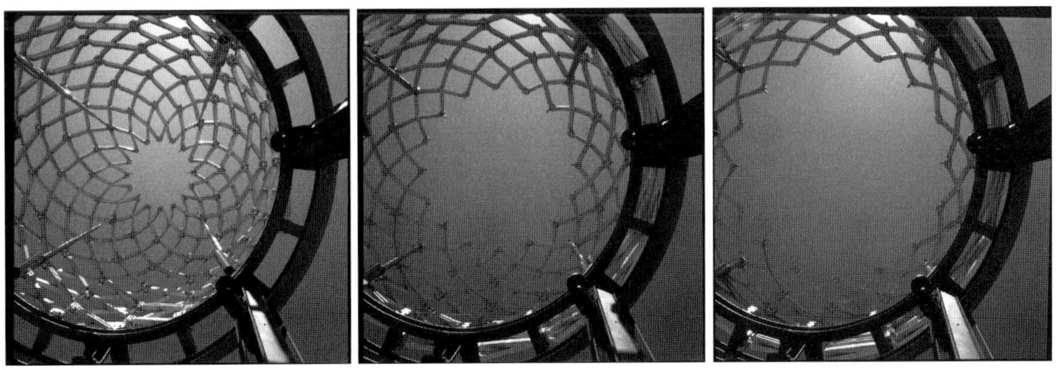

8.8 Iris Retractable Dome – the view from below

8.9 Mechanical Curtain

8.10 Movable panel construction

8.11 Raising the structure into position

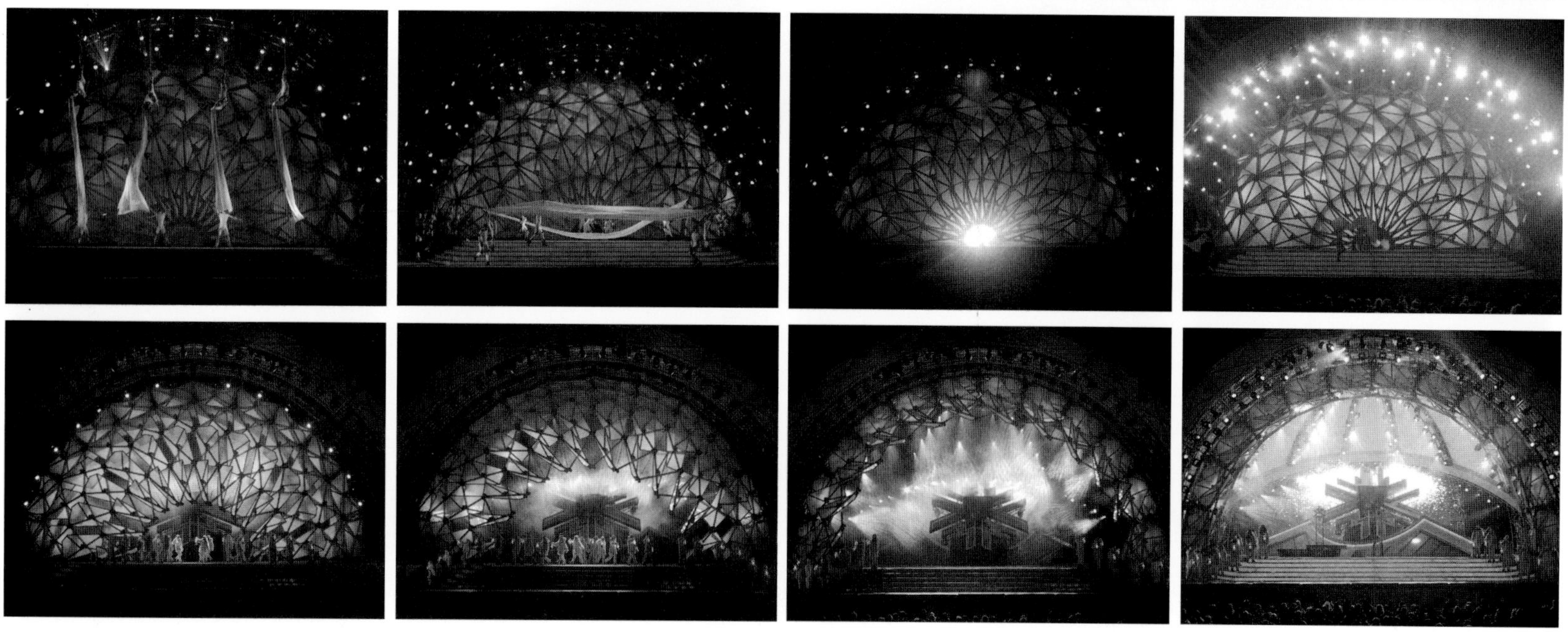

8.12 Olympic Games opening event

but which still enabled sufficiently light structural member sizes to be used to help keep the overall weight down. As the screen was orientated vertically, potential wind loads were a real concern, and because of its location within an open park these might be fairly significant. However, the transformation characteristics of the structure also proved valuable in this regard because if the wind speed ever exceeded 40 mph the screen could be retracted, and consequently it could be designed for a reduced wind loading of 0.3 kN/m^2.

The deployment of the Mechanical Curtain was part of a performance that included music, lighting effects and dancers. Over 500 computer-controlled lights were integrated into the arch's movements so that it entirely changed its colour and appearance as it opened (Figure 8.12).

Transforming architecture

For me, buildings are the ultimate interactive experience. We are immersed in them, moving through spaces with new vistas and perspectives constantly opening before us. My vision of transforming architecture extends and inverts this process: instead of moving through a building, the viewer may be stationary and it is the building itself that changes. Walls disappear and reappear. A roof spirals open. The inside becomes outside. An object expands and envelops one (Figure 8.13).

Potential applications for transformable architecture include; multiple-use spaces, spaces that alternate between indoors and outdoors, portable structures and shelters, retractable walls, sunshades and windbreaks, buildings that respond to climate change, interactive spaces for entertainment and many other facilities where flexibility and adaptability are required. The technology to achieve these possibilities is practical and ready for use.

8.13 Design for a retractable auditorium roof

Traces: The Architecture of Remembering

Sarah Bonnemaison
Dalhousie University

When X-rays were first invented, wealthy women commissioned radiographs of their bejeweled hand as a "self-portrait." The jewelry conferred status, bringing the novel image out of the realm of science and placing it firmly in the world of art. This interesting anecdote shows a fundamental impulse of the mind to navigate between science and art. On one hand, art – in the form of decorative rings – tames the frightening and even repulsive character of such science while, on the other, scientific inventions such as radiography make the world visible in new ways. The philosopher Michel Serres has explored intellectual questions that engage science and art simultaneously in his book *Le Passage du nord-ouest* (Serres and Hermes, 1980). For Serres, the floating icebergs and uncertain archipelago of the trans-continental route through northern Canada serves as a metaphor for investigation and discovery, as it has oscillated between scientific inventions and philosophical inquiry throughout history.

The subject of this chapter is informed by the interplay between art and science. The intent of the larger research context and the work undertaken within the design partnership, Filum, and with partner Christine Macy, is to develop an esthetic for tensile architecture based on movements, in particular the movements of dance.

Modernism is replete with art works which have attempted to depict the human body in movement, Umberto Boccioni's *Unique Form of a Continuous Space* (1913) and Jane Callaghan and Caroline Palmer's *Space Shapers* (1943) to name a few. In painting, Marcel Duchamp's *Nude Descending the Staircase* (1912) is the best-known early example of this interest, as well as Wassily Kandinsky's *Movement of Paluca* (1925). In the realm of science, Eadweard Muybridge's well-known stop-action photographs of animals and people in motion (1870) were followed by Thomas Eakins' "action photos" (1880), Etienne Marey's chronophotographs (1890), and finally Harold Edgerton's stroboscopic images (1950).

Certainly, modern architects' interest in the human body influenced their designs: from ergonomic and Taylorist studies to the doctrine of functionalism. At its worst, this work can be reductive and universalizing. Yet, beyond the simple work triangles and movement diagrams that can be easily construed too rigidly, some architects and designers have pushed their explorations of movement into new ways of seeing and inhabiting space. Innovative design projects based on analyses of movement include Lily Reich's kitchen designed for the Deutsche Bauaustellung in Berlin (1931) as well as the "choreographic architecture" of Eileen Gray, and her furniture designs that were entirely conceived in relation to the body in movement. The "architectural promenades" of Le Corbusier, the glass house of Pierre Chareau and the "Endless house" of Frederick Kiesler are other examples.

The work presented in this chapter revisits modernist interest in motion studies and extends it, using new technologies and a new esthetic. The new technologies of computer graphics and digital video extend our ability to visualize the human body in motion, recording and understanding the interface between movement and form in new ways. Technological innovations in the design and materials of tensioned structures along with the new technologies of computer-aided design and manufacturing have made it possible for designers to explore complex curvatures in architectural forms. The research also explores a new esthetic that has arisen from the contemporary renewed interest in incorporating a sense of movement in architecture. Spurred on by Frank Gehry's path-breaking work for the Guggenheim in Bilbao, and the potential of computer programs to generate the curvaceous "blob-like" forms of Asymptote and Greg Lynn, many contemporary architects are once again criticizing the modernist box and exploring alternative, and more expressive forms. This work aims to anchor such exploration in a theoretical and methodological position that sets the human body in the center of the logic that produces these forms.

The first part of the research was done in collaboration with choreographers Sandra Loring and Jill Ann Schwartz in 2003. While it is possible to analyze the everyday movements of people performing tasks at home or at work, the interest in dancers was fueled by their profound understanding of the way their bodies move and what those movements mean. In this collaboration, we hope to develop a more sophisticated and complex approach to design that goes beyond function into the realm of representation and poetics. The work illustrated in this chapter shows the results of research carried out through several seminars and workshops at Dalhousie University in winter 2004. These explorations concentrate on consecutively tracing the displacement of a dancer in space, translating these traces into meaningful visual data and transforming the results of the visual data into three-dimensional volumes of tensile architecture.

Tracing movement

How could one trace the displacement of the dancer's body in space? Traces left by footprints on a sandy beach, the sinus line by the ice skater, and the marks of our shoes in fresh snow, show that the weight of our body leaves traces on the ground as we move through space. These traces have fascinated artists and philosophers since ancient times, but during the nineteenth century the body movements of the worker also became the focus of research. Timing and tracing the repetitive movements of industrial laborers and housewives became a scientific endeavor. Space-time studies dissected movements requiring skill and craft into phases so as to reveal their inner structure.

Frank Gilbreth (who trained as a bricklayer) with his wife, the psychologist Lillian Gilbreth, developed analytical methods to study workers' movements and presented their findings in striking photographs (Giedeon, 1948). They disregarded the stopwatch of Frederick Taylor and replaced it with a recording apparatus called the "cyclograph" – an ordinary camera photographing small electric bulbs fastened onto the limbs of the person performing the task under observation. These tasks range from a surgeon tying a knot, a maid folding a handkerchief, a mason laying bricks to a fencer. The form of the movement, invisible to the naked eye, could now be recorded as a pattern of light on a single photographic plate. This was an important technical innovation over Muybridge's photographs, which segmented movement into a series of discrete images.

The cyclographs can also be seen as mysterious calligraphic forms endowed with symbolic meaning. Many artists were fascinated with these scientific studies of motion patterns. Paul Klee, one of the modern painters who could bring the greatest meaning, humor, and emotion to a simple line, said that "pictorial art springs from *movement,* is in itself interrupted *motion* and is conceived as *motion*" (Giedeon, 1948).

9.1 Movement with flashlights

9.2 Sand markings

9.3 Modern dance movement by dancer Maria Osende

Tracing with light

In an effort to trace the movement of one point on the body of the dancer, the intention evolved into creating a light path that is both beautiful and suggestive – an artistic object in its own right. With flashlights attached to their arms, each student performed a movement sentence he or she invented, following Klee's quotation that "[t]he performer is a point in transit" (Giedeon, 1948). Slow shutter speed recorded the light path on photographic film. The casual traces created by someone dancing freely contrast with the precision of another student executing a movement of karate or the circular pattern of belly dancing.

Like the brush of the calligrapher, the light path reveals the speed of movement – as the hand moves fast, the line is thin, but when it slows down, it thickens. A dot created by an instant of hesitation becomes a stain of light as the hand remains still for a longer period (Figure 9.1). These light patterns also reveal the psychology of the performer in the poses, hesitations, and diversions. In other words, "light tracings" reveal perfection in the gesture but also embrace sources of emotion.

Tracing with sand

Sand was also used to record the displacement of dancers' feet on the ground. Using a funnel filled with sand, an architecture student followed on the dancers' heels, drawing a line that traced the exact displacement of the feet on the ground. Tinting the sand with colored chalk helped distinguish segments of the choreography – especially useful when different paths overlap one another (Figure 9.2). One such sand painting follows a radial pattern, second is a series of interweaving curves and the third one recalls a spiral. The sand is carefully brushed aside and replaced by chalk to create a permanent record of the sand patterns. Creating large colorful images on the ground, the chalk is then replaced by paint. Like the light tracings, the sand paintings reveal a great deal about the emotions and intentions of the dancer. These sand patterns share qualities with Jackson Pollock's drip-paintings – which are a rich hybrid of abstraction and emotion.

Sand tracings translate the choreography at the ground plane, and light tracings evoke the quality of the movement in the vertical dimension. However, both flatten the movement onto horizontal or vertical planes. Following these studies, the need to find another mode of analysis to deconstruct the three-dimensional aspect of the choreography became evident.

Projecting shadows

Shadow projection is used as a way to transcribe the volume of a dancer onto planes, i.e. plan and elevation. When the planes (or the projection surfaces) are placed at varying distances from the dancer and the light source, the movement of the dancer is deconstructed in many different ways (Figure 9.3). This part of the work was done during a seminar at Dalhousie University during the winter of 2004.

By placing one light in front of a dancer and another above, shadows were cast on both the wall and the floor, giving the equivalent of an elevation and plan view of the dancer. These shadows recall the allegory of Plato's cave, transposed to paper screens. The contours of these reflected shadows were drawn directly onto paper screens. Each pose led to one contour and the final drawing was created by a multitude of contours overlapping one another. The contour drawing does not trace the displacement of one point in space (as did the light tracings), but is a projection of the whole surface of the body. In other words, this technique gave us a record of the shape of the whole, rather than the movement of a point. The overlapping contours recall the staccato effect of movement under strobe lighting as the imagination fills in the gap between each contour and generates a sense of movement. As an esthetic and set of conventions, these contour drawings can be seen as deconstructing and saving the body at the same time. The next step deals with how we transform these visual data into a three-dimensional structure that speaks of that movement.

Wire models

The Gilbreths translated cyclograph images into models constructed out of wire in order to show the path of the hands in a more accessible way. "These wire curves, their windings, their sinuousities, show exactly how the action was carried out" (Giedeon, 1948, p. 104). These models show where the hand faltered and where it performed its task without hesitation. They were meant to help the participants to look at themselves objectively. In fact, as Giedeon (1948, p. 106) explains, "the worker, able to see his own gesture in space-time representation, should become what the Gilbreths call 'motion minded'." In this case, we see wire models of a bricklayer. By understanding when and where energy was being wasted, it was also possible to introduce props (such as scaffolding to store bricks) to reduce fatigue in the worker's body.

We used wire to model the choreography in an effort to record the path of movement as a three-dimensional representation, and to reveal an iconography that led to generation of new forms. The sand tracings described earlier were revisited to model the three-dimensional aspects of the choreographies. For example, the choreography that produced the sand painting most like a spiral now generated a relatively enclosed space like a cocoon. For the dancer, the wire model – like the path of light – became a sort of mirror, reflecting the beauty of the movement in its inner form, pulsation, and even desire. In that way, the tracery of light and the wire models reveal the motion in its full plasticity. Motion acquires a form of its own and a life of its own.

In order to find the tensile surfaces generated by each wire model, we dipped them in a soap film solution. The soap film settles in the least stressful configuration and gives us a stable geometry within the wire structure. We can say that it is a process of form-finding as opposed to form-making. If the model defines an enclosed volume, the soap film membrane tends towards a crystalline geometry made of flat planes. But if the wire model has a lot of curves and an open-ended geometry, the soap film stretches into sweeping double curves.

Transformation

If we look at the work of modernist dancers, we see shared affinities between our models of dance movements and their early explorations into the geometry of movement. The double curves created by the soap film models recall the sweeping curves of silk manipulated by Loïe Fuller in the early part of the century. In her *Danse du Lys* of 1902, she bathed her fabrics with luminescent salts that responded to the gels in the theatrical stage lights, creating shimmering iridescent bubbles of form in movement. Stéphane Mallarmé, the Symbolist poet, described her dancing as "the dizziness of soul made visible by an artifice."

On the other hand, the crystalline forms of the enclosed wire models recall the more analytical geometries of Rudolf Laban. Laban wanted to place the body of the dancer in a grid of platonic geometry that he called a *Kinetogramme* (1910). Using this spatial geometry, he developed a grammar of gestures in which the limbs would reach the distant points of the polygon called scales. In these scales, he demonstrates how body movements are broken down into axes and orientations. While both Fuller and Laban worked with scientific ideas, Fuller's research was oriented toward chemistry and light, and Laban's toward creating a grammar of dance that has endured as the only dance notation in use today.

9.4 Tensile model from classical dance by Caroline Caskey

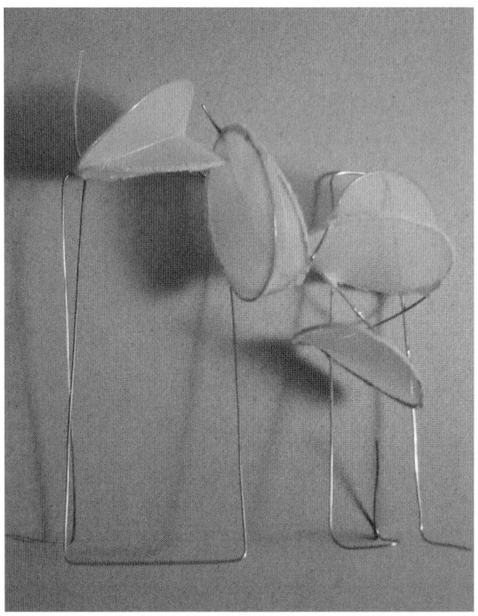

9.5 Tensile model from modern dance by Jane Abbott

Transforming the shadows

Having revisited the work of Rudolf Laban and Loïe Fuller, we now embark on creating three-dimensional transformations of the shadows. Linking points from the vertical and horizontal shadows, we can generate a series of volumetric studies. A series of images describe a complete turn on one leg. The forms inhabit the same space as the dancer's body but they are of a radically different nature. These volumes are no longer pure *translations* of dance movements, they are *transformations* of two planes into three-dimensional shapes. A complete turn performed by Maria Osende with one leg extended results in a three-dimensional tensile model that recalls a cocktail glass (Figure 9.4). Like the early X-ray images of the human body (Cartwright, 1995), we are "making darkness visible," metaphorically and literally. Metaphorically, we are isolating vectors of force, points of balance, rotation, and extensions within the dancer's body. Literally, we are making darkness visible by piercing the opaque plane of the shadow.

Interestingly, the models generated from classical ballet are radically different from the ones generated by modern dance. For example, the model depicted in Figure 9.5 is a transformation of the modern dance seen in Figure 9.4. Classical poses lead to forms structured by vectors anchored on a strong central core. The trunk stays straight and the limbs project outward. This has an affinity to the work of Oskar Schlemmer in the 1920s. In his preoccupation with space, abstraction, movement, and light, Schlemmer staged a series of experiments that dealt with the "stereometry of space," and the effect of the abstracted human figure on that space. In *Pole Dance*, for example, the function of the slats strapped to the body of the dancer was to outline the geometrical division of the space occupied by the dancer and emphasize the perspective view for the audience. As in the work of Laban, the body is firmly centered and all geometrical lines radiate outward from that center point located at the chest.

By contrast, modern poses lead to pure volumes interlocking in one another where the body carves out forms in which a constant tension between the positive and negative spaces is created. This has an affinity with the sculptures of Henry Moore and Antoine Pevsner such as *Projection dans l'espace* (1938–39). In both cases, the object makes a clear distinction between tensile and compressive elements and carves out the space with a rigor that is nearly mathematical. But, whether we look at the transformations generated from classical or from modern dance, the models reach deep into the essence of the dance style from which it is generated.

Building full scale

In order to build the full-scale structures evolving from the modeling stage, it is necessary to close the loop between the movement, the translation, and the reinterpretation of the space by the dancers. Students first constructed the spiral-like form out of small trees bent into shape. The built space is like a cocoon in which the soap film surface has been replaced with fishnet and the sand painting has been transcribed with a black line on the platform (Figure 9.6) which is just large enough for the dancer to move in.

The sand painting made of curved strokes on the ground led to a series of small circles linked by a continuous line. This was built as a large *ikebana* arrangement, with suspended hoops woven out of vines forming each circle. These hoops reacted to the dancer and the dancer to them.

The wire model with large outer circles and small inner ones became a series of fabric elements describing a cartwheel, a major movement of the choreography. The performance structures that were built around each of these three dance solos were developed as spatial translations of each dancer's original choreography. Performing in these structures developed out of their own movements, the dancers then had to modify their choreographies, adding new movements in response to the new setting. In this play of action and reaction, the relationship between the built space and the dancer was like an embrace: each retained their identity but the result was larger and more powerful than if either had been alone.

Conclusion

There are two more phases to this research project, which are supported by a Research/Creation Grant from the Social Sciences and Humanities Research Council of Canada for 2004 to 2006. The first one includes tracing movements with the help of a motion capture stage and translating the data into three-dimensional digital models. We will then manipulate these models to build an environment for the choreographers to investigate. The second stage of the research involves creating an environment that is responsive to touch, and again investigates the choreographic potential of such spaces.

9.6 Tensile structure by Lefoko Simako and Michelle Yeung

References

Cartwright, L. (1995) *Screening the Bod: Tracing Medicine's Visual Culture.* Minneapolis: University of Minnesota Press.

Giedeon, S. (1948) *Mechanization Takes Command: A Contribution to Anonymous History.* New York: Norton.

Serres, M. and Hermes, V. (1980) *Le passage du nord-ouest*. Paris: Les Editions de Minuit.

Plastic and Bamboo: Tailor-made Tent Design

Marcin W. Padlewski

Introduction

Tents are adaptable structures that are lightweight, transportable and temporary. A variety of existing tent forms such as the African black tent and the Asian yurt are built with available materials. These traditional tents respond to the demands of a nomadic lifestyle by evolving naturally with climate changes. Contemporary practice in tent design also has the ultimate objective of responding to the desires and necessities of life quickly and efficiently with available and affordable materials.

The tent designs described in this chapter were built between 2001 and 2004, in Ottawa, Canada, by Marcin Padlewski, Michel DuVernet and Anissa Szeto of design team, the Bakery Group. These design proposals differ from most contemporary precedents as they were all explored by using a *hand-made* approach, without computer-aided design or engineering analysis. In essence, the method used encouraged building processes and the use of tensile and deployable structures simply and instinctively. This chapter presents the methods and processes used in the construction of these prototypical 'tailored' tents and other deployable structures utilizing plastic and bamboo.

The area of emergency relief housing requires further design investigation to improve the available options for the provision of basic shelter. Emergency conditions can be caused by a wide range of natural or man-made disasters including earthquakes, landslides and political conflicts. Major relief organizations supply tents or tarpaulins, rectangular sheets of plastic, which are cheap and immediate solutions to shelter people in need. The inadequacy of this relief effort system becomes evident through accounts of camp situations that identify families huddled and exposed under these sheets for weeks after initial help arrives. Sometimes, as a second wave of relief efforts, traditional canvas tents are supplied. Although these are heavier and more expensive to transport, they remain difficult to set up and problematic to maintain. Larger tents, commonly used for mid- to long-term solutions, are usually insufficient in quantity to provide housing for all those in need.

In an effort to provide a better and more efficient emergency shelter for use in conditions caused by short- to mid-term humanitarian crisis, Bakery Group's design strategy combined the use of inexpensive materials with simple fabrication techniques. The following tent designs evolved in spite of the lack of support within the seemingly intricate and bureaucratic workings of the humanitarian relief market. Exploring emerging ideas for deployable architectural forms, our emphasis evolved towards the simple manipulation of the materials and the structure used. The results of the iterative-evaluative design process informed the construction of the prototypical tent design throughout the process.

Geometry, materials and method

In order to achieve a shelter design that is cost effective, our method combined the use of inexpensive materials with minimum assembly requirements in the production process. We also believed that if a shelter could be built and assembled easily and effectively by hand, then mass production would also be quick and easy. To provide a stable tent skin, our initial assumption was to integrate a *hyperbolic paraboloid* or a saddle surface into the design of the tent's canopy. The saddle shape was stabilized by applying tension to at least four points in two opposite directions. The black tents of Africa and the Middle East, very stable in adverse weather conditions though not easily deployable by a non-native, are among the oldest examples that use this structural method. Our search for a system that could be used, assembled, dismantled and deployed easily by anyone continued.

We chose reinforced polyethylene film as a canopy material as it is low-cost, light-weight, tear-resistant and fire-retardant. Polyethylene can be easily heat-welded using metal bars or a hot air jet. We have successfully bonded this plastic material using high-pressure steam. Polyethylene film is an isotropic material that provides equal strength in all directions. In order to create a saddle surface, it had to be cut and assembled from individual parts, which is a potentially time-consuming process. However, we discovered that a single flat sheet could be manipulated into a three-dimensional saddle surface by what we called the *pinching* technique. Original experiments in pinching used simple, unframed membranes with individual tension points. Later trials have shown that an integrated frame or a continuous edge is more advantageous in transmitting the tensile forces from the tent's surface. Although the frame is not structurally necessary, it serves to effectively stabilize, deploy and keep the canopy in tension. We had chosen Tonkin bamboo as the frame material for its properties of natural elasticity, light weight and sustainability. The frame consisted of four poles attached together into a closed loop that spread apart at the peak and base. The canopy was mounted within the straight-edge frame which resulted in a highly stable yet collapsible structure.

Pinching

The tent frame is pre-assembled from four straight bamboo pieces coupled with four flexible (hinged) elbows. A single piece of polyethylene film is cut from a continuous roll and heat sealed onto a hinged frame. The frame and film are then folded and pinched, i.e. sewn or welded together. In one step, two overlapping layers of film are welded and cut along a curved line, thus producing a watertight seam. The remaining crescent is then removed. Next, the frame and film are opened and refolded perpendicular to the first seam. A secondary pinch is applied and the crescents are removed. The frame is once again opened and kept in tension at the peak to produce a double-curved or saddle-surface canopy (Figure 10.1).

Tailoring saddle surfaces is achieved by pinching curved seams along the folds of a tent's surface. A surface bound by a frame can only be folded at the hinges of that frame. A curved pinch along this fold breaks the flat surfaces and produces two curved surfaces joined by a parabolic seam. Folding the curved surfaces perpendicular to the first seam and applying a second pinch results in a saddle shape. Spatially, the result is a set of two opposing curves crossing at their vertex at right angles. This demonstrates that this simple method can be used to create complex surfaces by hand. Additionally, the shape and proportion of the saddles produced in this manner can be adjusted by controlling the pinching curves and frame proportions. The saddle shape is controlled and manipulated into the desired form by changing the depth and shape of each pinch. Only the aperture of the frame limits the amount of variable saddles. To date, our explorations show that a six-sided frame would require three pinches, while a four-sided panel represented the simplest double-curved surface made in this manner, and a three-sided surface resulted in a flat plane.

10.1 Pinching process

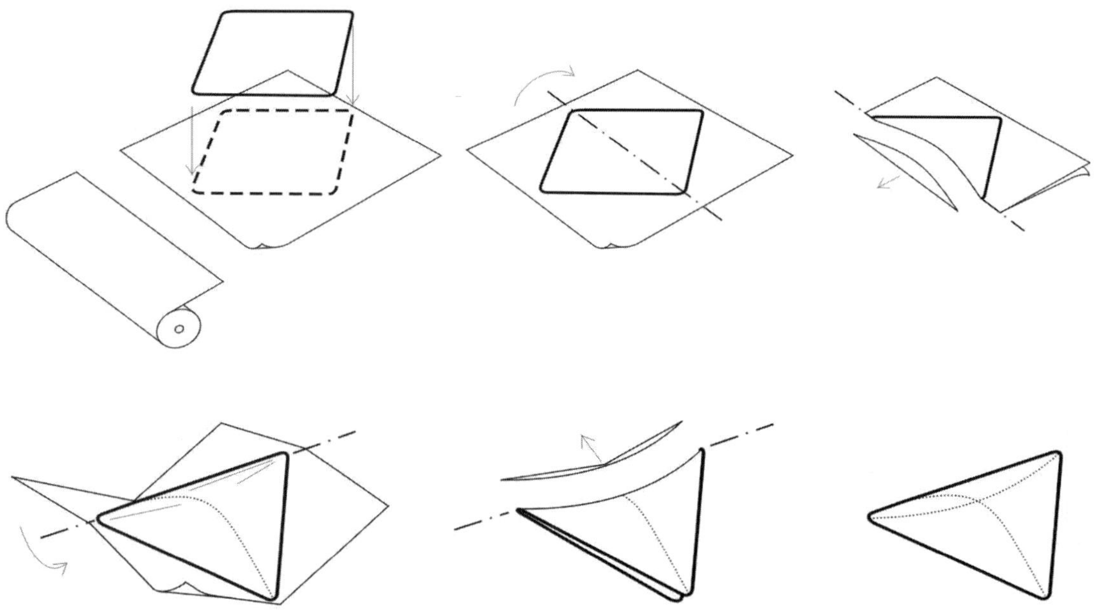

10.2 The *basic shelter* tent deployed and folded

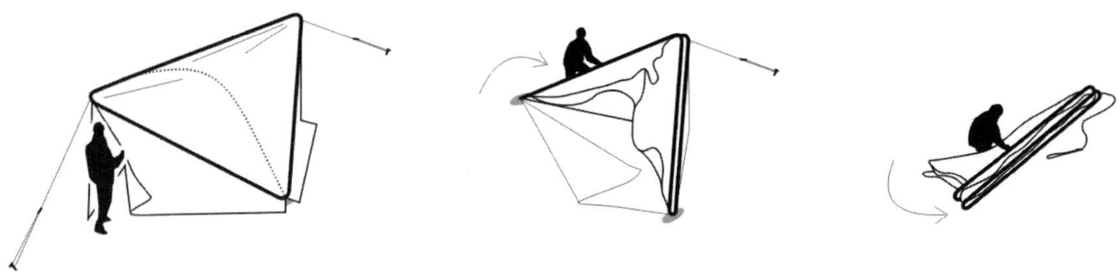

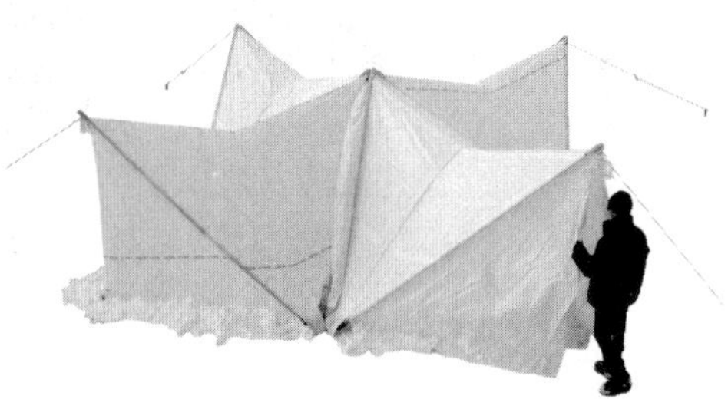

10.3 Four interconnected *basic shelters*

10.4 Deployment of a *tent-panel*

Basic shelter

Our proposal for a basic emergency shelter, made up of a square plan with dimensions of 2.3 by 2.3 by 2 metres high (7.5 by 7.5 by 6.5 feet), four vertical walls, and a large framed canopy, employs the pinching method. The tent's canopy is effectively kept in tension by the open frame and two outer guy lines. The walls hang free and are weighed down by available material such as sand, rocks or snow. The floor surface is attached to the bottom joints of the frame which determines the tent's maximum spread. Given the scale of this design, providing a single pinch across the tent's canopy is enough to create a simple curved space with an effective structure (Figure 10.2). The square plan and elevation of a single tent allow it to connect with other adjacent tents, creating modular building blocks that are capable of creating endless ground architecture for use as a family dwelling. Flexibility of layout and size creates spaces for multiple functions. The tensioning of each tent module in larger assemblies is achieved by latching one frame to the next, an advantage of the straight frame (Figure 10.3). Linear tunnel-like connections are found to be the most efficient way of linking the modules.

The first set-up of a *basic shelter* module, including its deployment and securing the structure to the ground, was done in less than one minute. This first version was put through a full year of outdoor exposure, enduring the high winds, ice and snow of an eastern Canadian winter. The canopy – the pinched membrane – was stable throughout the year. But since climate and conditions where a tent may be set up can vary, the tent's skin is adjusted for seasonal temperature changes with the simple addition of insect netting for airflow and a synthetic felt for insulation.

In 2002, the basic shelter was exhibited at the International Aid and Trade Show in Geneva that is attended by various relief organizations and manufacturers. Although there was interest in the basic shelter design due to its ease of deployment, this simple and stable design awaits its true application. Its efficiency as a low-cost, rapidly deployable emergency shelter remains to be tested in real-life situations. However, basic shelter was the modular building block that formed a foundation of all further design explorations.

The tent-panel

To expand on other tent prototypes, we focused on the framed canopy, i.e. the roof portion of basic shelter and the principal tensile component. Up to this point we had relied on the ground plane to anchor and keep the tent stable – this proved to be a limiting factor. For the next step, a freestanding structure was created by adding two compressed struts onto a four-sided panel. The struts, located tangent to each side of the frame and across the centre of the double-curved membrane without any protrusion beyond the curvature of the surface, allowed a true self-supporting structure. The tent-panel became a self-tensioning, collapsible and hyperbolic building block (Figure 10.4). The self-supporting tent-panel can be individually used as a shading device and shelter. It is also feasible to create large free-form installations by connecting one panel to another of varying proportions and openness. Whole assemblies can be mounted over or onto buildings, creating potentially interesting organic undulations (Figure 10.5).

Panels and tunnels

Following this experiment, we proceeded to apply the tent-panels (or cells) as repetitive elements within one continuous skin. By controlling the number of panels and their scale, a balanced structure and covered space was achieved. Tunnel tents and blanket structures can be made in a similar way to the basic shelter without repeating multiple pinches that generate cells in a continuous and

10.5 Simple *tent-panel* assemblies used without struts

uncut surface. By staggering and overlapping the cells, an efficient spatial enclosure can be created. Tunnel tents – simple to make and easy to use – offer free-standing design alternatives that are deployable with minimal assembly and handling requirements. The free-standing accordion tunnel tent reacts slightly to wind forces, and yet always returns to a central balanced point due to the elastic properties of the bamboo and jointing methods (Figure 10.6).

Although the work explores mostly symmetrical designs, pinching multiple cells in a continuous membrane is also possible for creating structural covers (blankets) with varying shapes. Covers produced in this manner can take on many forms and still remain structurally firm. Their outer profile can be made to resemble irregular shapes such as cloud formations. This again can be achieved by controlling the size and proportion of each cell. The tent-panel, a cellular and geometrically rigid component as part of a larger structure, creates a tensile space frame.

Full circle

The assembly of tent-panels into a concentric and circular plan resulted in a scissor structure that was generated by rotating the frames of the tent-panel and pinning their intersections. A central hub was added to connect the frames at the apex. The number of elements was increased to generate a sixteen-sided polygon, approximating a circular plan. The deployment of this structure is quite simple (even more so than the traditional Asian yurt which it resembles) because of the pre-assembled framework. The lattice wall is combined with the roof poles, acting together as a giant umbrella with side walls. Once the structure is unfolded, the cover is simply thrown on and stretched down tight. The canopy does not require pinching in order to work as a cover. In a larger version, however, the roof to wall transition would probably benefit from saddle-shaped surfaces made of bamboo and reinforced polyethylene plastic. This first-generation collapsible yurt-type dwelling is 4.3 metres (14 foot) in diameter and 2.6 metres (8.5 foot) high, and weighs approximately 24 kg (53 lbs) (Figure 10.7).

Process-driven design

From saddles to circles, these tent design prototypes are the result of a process driven by materials and built experiments. The simple pinching technique proves to be instrumental in allowing us to explore built forms quickly and efficiently without the need for special tools for assembly. Although machines can be used in manufacturing these tents for speed, plastic and bamboo can also be sealed and joined by hand. Adding off-the-shelf building supplies and other simple material combinations can often lead to unexpected results. As our design process still unfolds, there is much more to explore in the simple act of building a tent.

10.6 Integrated multi-celled accordion tent

10.7 Deployable yurt structure

Pedestrian Clip-on Footbridge: Making Use of Temporary City Space

Andrew Furman
Ryerson University

Introduction

In the complex density of cities innovative artists and designers must explore alternative strategies for finding spaces for urban interventions. One such strategy concerns itself with introducing or transporting urban elements (for example unexpected densities or program functions) into rural or suburban sites. Another strategy exploits the body as a transportable device relying on its own means of locomotion to enter *wilfully* into a host site. Installation artists and designers may also take advantage of the transitional phases of an urban site in city centers for decades. City development projects and construction schedules do not typically offer public use of a site between the end of its prior use and the substantial completion of the latest refurbishment/new construction, safety issues being a major concern for not allowing the public to enter an "unfinished" site. However, there have been some successful temporary events and projects that have take place during what is typically considered this unoccupiable period in the site's history.

The discovery of desirable space for a temporary installation or program that takes advantage of a noticeable entry point into a building via the street (or some other visible public access point) by pedestrians is something that occurs on a trajectory between the undeclared and the declared use of a site. Connecting to a temporary space may be via a cryptic entry sequence or a clearly demarcated public route. Radically different approaches to the construction of a transitional space may determine how a temporary program can mobilize people. For example, a model suite/sales venue within a working construction site, cordoned off from the overall construction activity, suggests a very different pedestrian encounter than, for example, chancing upon the interventions in an abandoned site by Gordon Matta-Clark.

A photographic record of an abandoned building may powerfully demonstrate the idea that Roland Barthes coined in *Camera Lucida* (1998): the *punctum*, the personal and emotional response a viewer has to an image, the detail that one takes notice of that also makes the space/event different from the normative. By physically linking a site to the city, the installation artist/designer allows pedestrian visitors to experience urbanity as they move and discover spaces through chance, distraction, desire, and opportunity.

Ramps, bridges, and construction

Increased demands for maximized movement of motorized vehicles and marginalization of the pedestrian use of streets are known facts in most city centers. Exceptions to this are the pedestrianized areas for commerce and tourism that act as a counterpoint to unimpeded traffic flows. Somewhere between these two extremes, we may find responses to the conundrum of providing enough street space for both pedestrians and vehicles. Specific responses to the constraints of city space design have in some cases led to dramatic engineered solutions such as physically elevating or excavating the street for ease of pedestrian or vehicle flow.

Installations and temporary events may borrow from some of the principles of city street engineering, and also some assumptions from construction practice. The extension of the urban street has been developed in transportation planning with two dominant ideas – one is the maximization of vehicle density (parking) and the other is its dispersal (highways). Physical depressions of the street such as ramps leading to underground parking structures are often visible right in the city center. Open-air highways and ramps that utilize posts or earth berms to exit or enter city traffic are visible at the fringe of the central core.

During a building construction, hoardings and covered walkways transform sidewalks and roads; cranes cross public and private airspace; streets are closed and traffic is redirected; city parts usually hidden are revealed; and all manner of materials are conveyed to and from a site, often crossing dominant pedestrian routes (Figure 11.1). Of course, these conditions are not directly concerned with the pedestrian. However, the adoption of the language of construction ramps and bridges, often emphasizing the pedestrian crossing over/into a construction site or something crossing over the pedestrian route, suits many temporary installations well.

Temporary programs

Gordon Matta-Clark's large sculptural intervention *Day's End* was located in an abandoned pier in Lower Manhattan in August 1975. In it he carved out and removed materials from various parts of the host building. His work was experienced simultaneously through the new visual pattern of light introduced by the cut-outs, and also the new walking path that required special attention whenever crossing a voided area (for example he introduced a wooden plank over a voided floor, revealing the water of the pier beneath). The site was secured against further intruders, and after two months of steady work, the project was ready for visitors. The authorities closed down the installation soon after and Matta-Clark retaliated by insisting that the work was part of the "public domain." He further suggested that it represented an artistic contribution to a "decaying city" (Lee, 2000, pp. 122–123).

Installation artist Tadashi Kawamata incorporates new construction or salvaged materials from demolition sites in his projects that become resources for future installations (Joseph and Look, 2003, p. 47). He has an interest in utilizing vacant and abandoned sites, open squares, and historic interiors; the massive wooden constructions often incorporate elements of a bridge or walkway, and are accessible to the public. His installations slowly transform into a space that may be inhabited by the pedestrian – they develop in a similar way to a typical construction site, and may appear at first glance to be a chaotic wooden hoarding. In *Colonial Tavern Park*, a project undertaken in the late summer of 1989 in Toronto, the wooden scaffolding he created lay between two historic bank buildings, in the void of a recently demolished jazz bar, now a park space (Figure 11.2). Kawamata's work reflects his concerns for the public perception of modified/renovated environments. Bettina Paust writes that often "the public pays attention to the beginning or to the end of an event and doesn't see the in-between, the time which forms, the time of conception" (Joseph and Look, 2003, p. 44).

Above Times Square (1999), a project by Lot-ek with Mark Robbins, was a proposal concerned with the physical lack of pedestrian space for tourists as well as inhabitants in Times Square, Manhattan, New York. An elevated catwalk, intended primarily for tourists, was stacked above street level and created a linear pattern for viewing spaces in the center of the traffic. This extensive bridge could "be variously programmed, changing to accommodate larger crowds and the everyday viewing of daily traffic in the square" (LOT-EK website, 2004). It connected vertically to the framed and illuminated

11.1 A typical construction site at the sidewalk in Canada

11.2 Kawamata's site in Toronto under construction (courtesy Mercer Union, Toronto)

11.3 *Above Times Square*, project by LOT-EK and Mark Robbins

11.4 The footbridge in context

billboards (Figure 11.3) that line the canyon of Times Square. Creating a vantage point from which to view and experience the entire streetscape in an altogether new way, pedestrians could move through various levels around and in the billboards.

Project background

The project described here was installed during the first week of an architectural event in Munich in the summer of 2002 and it demonstrates that vacant and leftover spaces are potentially rich opportunities for pedestrian-focused temporary programs. As part of the pre-conference events that supported the main Union of International Architects (UIA) conference in Berlin, the local and regional design community wished to find a suitable site for the main lecture and event space. Initially, none of available spaces suited the requirements for opening and closing ceremonies. Eventually a space was found that was large enough to accommodate both functions. The footbridge that formed the entrance to this space was commissioned by the Bavarian branch of the Bund Deutscher Architekten. This temporary transportable structure was realized through a collaborative effort made up of three local Bavarian architects: Thomas Beck (engineering phase), Matthias Castorph (initial conception), and Peter Haimerl (planning and execution phases). It was installed for a period of ten days over the Dienerstraße Street between the Alter Hof's west side and the northeast side of the Marienhof Park in Munich (Figure 11.4). Running concurrently with the final three days of the main event, the World Congress of Architecture (UIA), Munich pre-conference was held on July 19–21, 2002.

The architects had to deal with a difficult existing spatial sequence; a second-story space; an interior with repetitive punched glazing behind a nondescript exterior facade, a long rectangular volume located in the northwestern portion of the Alter Hof complex, the "oldest and only purely medieval building in modern Munich" (Wadleigh, 1910) (Figure 11.5). The footbridge served as a visual marker for the area, more powerful in its form and use than the ostentatiously placed re-zoning signs, the first symbol of a redevelopment proposal. It was a physical representation of the collective memory and physical experience of a building slated for change through additional use and event spaces. The footbridge led into a long, rectangular-shaped main hall subdivided along its length by a temporary wall that adjoins a café/bar in the main entry area, and gallery/display rooms beyond. The second floor, called "Architekturklub," included a café; a space for lectures and performances; a gallery space for design proposals, student work, and sponsor product exhibitions; and a nightclub with celebrated DJs and entertainers.

According to Peter Haimerl, the footbridge is symbolic of a larger vision of how existing cities in Europe could develop because it is a transportation device grafted onto/into the existing urban fabric, which is commensurate with Haimerl's interests in urban planning and transportability. As part of a hypothetical project entitled *Zoom Town,* he proposes a reconfigured Europe-wide mega-city system, free of cars in the urban areas that incorporate various level changes. Within this future city, he has developed a prototype for a transportable and self-contained living/work unit. His concept for stacking "pioneer-spaces" onto the vertical supply sticks, i.e. towers, proposes horizontal connector platforms reminiscent of interchangeable stacked living quarters typified by the 1909 cartoon reproduced in *Delirious New York* (Koolhaas, 1984) and other design manifestos. According to Haimerl, this is a "freely disposable living space" that "will redefine roadways and urban form by providing the planners and inhabitants with an instrument that will enable the development of democratic processes."

The footbridge challenges the status quo of the street. It has precedents in the visionary ideas developed by Robert Moses in the 1960s in his mega and meta-city proposals, and more recently in projects that parallel some of Haimerl's concerns, such as Wing Maas, Jacob van Rijs and Nathalie de Vries' (MVRDV) *Metacity/Datatown* in 1999, and LOT-EK's Times Square elevated pedestrian bridge proposal.

The architects have maximized the opportunity available from a vacant space that was being planned for a substantial renovation and transformed it beyond the public expectations through a significant change of use and introduction of a new access point. The suddenness of its arrival at the site is characteristic of many projects that deal with the transportability of architecture.

The footbridge

The footbridge provided a rich experience for pedestrians. A change in the relationship of tread depth along the path up to the window opening defied the standard architectural practice. The rules for building staircases/ramps were appropriated for a heightened sense of drama throughout the ascent (Fitchen, 1989). The logic for the reversal of this ladder-like proportioning system arose out of the need to maintain an equal rise for the steps as they followed the curvature of the footbridge, leading to consistently longer treads. The experience of using the bridge and crossing the dominant street traffic caused a reaction that was "forcing the viewer to ponder objects that are so familiar that they are ignored" (Viso, 1996).

A clear span of 23.5 meters and maximum rise of 6.76 meters provided enough height to experience an elevated sense of view. Of particular interest was the largest step/platform that occurred at the mid-point of the Dienerstraße; this was a clear indication of its secondary use as a vantage point for pedestrians to observe the view while resting. This recalls the popular esthetic techniques of leaving out any construction on the bridge at the mid-point in order to afford a distant view. This technique was incorporated into the early design of bridges such as the Ponte Vecchio (Kostof, 1999, p. 40).

The pedestrian bridge, built from an engineered laminated wood structure and pre-assembled off-site at a factory, was brought to the site by truck with the stairs already fabricated and affixed to the arced form. The footbridge (without the metal railings) was lifted by crane over Dienerstraße, and secured at the windowsill from the Alter Hof, and also at the far end where it rested on the Marienhof Park. The structure needed to be introduced to the site with a minimum of road closures which led to the engineering of the span supported only at two connection points. In this respect, the footbridge also borrowed from the precedent of Scarpa's asymmetrical wooden footbridge that crosses the narrow canal from the campo to the lintel (window opening) of the foyer in the Fondazione Querini Stampalia in Venice (1963). It uses two different systems of connections, expressed by the different spaces that are traversed at either end of the footbridge (Frampton, 2001, p. 302).

Since the footbridge was made up of one main structural element, the windowsill and the ground connection had to be prepared prior to hoisting the laminated beam to minimize disruptions to the scheduled street closing. In the Alter Hof structure, large wooden blocking was introduced to take up some of the lateral loading at the main points of contact. Two splayed I-sections transferred the load at the sill and distributed them into the masonry walls of the structure. Bolted connections to wood and masonry were utilized. At the park side, the landing pad was prepared with a perforated steel plate. This secured the bridge into the earth with the aid of fifty 1-meter-long by 75 mm-diameter spikes that take up the diagonal forces. Thirteen triangular metal plates were welded perpendicularly onto the main plate to receive the arced form of the

11.5 The footbridge from the park side

11.6 Staircase portion of the footbridge

11.7 Looking up the stairs from park side

footbridge. It was secured with ten bolts each side to resist the vertical forces. After the two ends were successfully secured, a layer of gravel was placed over the 2 by 1 meter metal plate to smooth out the transition between the last step and the earth. Then the main posts for the large railing were bolted in place. The red corrugated skin was applied after the installation of the main arched railing. The actual bridge appears very light, as if it were a hollow metal construction, largely due to the metal corrugation.

Over 3,000 visitors used the bridge every day during the first week of the architectural event. The venue was open daily late into the evening for special events and a nightclub. Once a pedestrian made the ascent up the thirty-three steps (typical riser height of 19 cm) and over the slight curved portion, there were seven additional wooden steps that led down from the windowsill level to the existing second floor. The footbridge was approximately 1 meter wide at the steps near the base and fanned out to comfortably accommodate two people walking in opposing directions. Automated lights illuminated the steps in the evening for easy navigation.

The public were encouraged to become active participants in the Architekturklub's program and circulation strategy. Implicit in the ascent of the stair/ramp is a design strategy that transforms the private spaces of the adjacent buildings into an extension of the park space across the street below. The passer-by might ask the questions: Does this bridge have permission to be there? Is it permanent? The feeling it created was confusing: Is it a new permanent entry or is it being used temporarily for the purpose of moving construction materials? The ambiguity of the footbridge lay also in its materiality; a red corrugated skin (reminiscent of shipping containers) juxtaposed against plywood, a substrate awaiting a final (perhaps permanent) exterior coating. The bridge hovered above in a manner that neither disrupted the motor vehicle flow below, nor compromised the predominant circulation of the Marienhof Park. It could be also interpreted as the inevitable (infill) development of the Alter Hof which is sending out a probe into the park, testing its readiness for public acceptance (Figure 11.5).

The footbridge was apparently unguarded and allowed for a kinesthetic interaction with the facade of the building. Normally, building entrances function as a filter, preventing access to those "dangerous and unpredictable objects (people)" (Michael Balint, quoted in Hall, 1990, p. 60). The bridge, through its physical grafting into the Alter Hof and the park, questioned the parasitical relationship of host and guest. Was the bridge an extension of the interior space or was it an extension of the park? The corrugated ribbing (Figures 11.6 and 11.7) made it a difficult target for the graffiti marks, posters, or advertising scrims found in most hoardings, historic reconstructions, and scaffoldings.

Soon after the ten days of events in the city, the space was resealed in preparation for its impending construction. The expected renovation to the Alter Hof proposes conversion of the north and east portions of the complex into a mixture of shops, restaurants, residences, and cultural spaces reflective of the current energy in the surrounding area. The east and west portions are government-owned and conform to the existing rooflines of adjacent buildings. Significant fenestration will be retained to adhere to Munich's historic preservation bylaws (Thalgott, 1999).

Conclusion

This footbridge was a temporary structure intended for use over a ten-day event. However it was later transferred to a park outside Munich, where it remains as a landscape structure spanning a small body of water. In this new location it serves as a reminder of the modern objects that define our experience, and the nature of our engagement with the street in the urban realm.

As the private sphere of built environments grows exponentially with the augmentation of communications and movement technologies, strategies within the urban core focus on the shrinking public spaces that are accessible to all. As in the example of the footbridge, small-scale temporary and transportable projects inserted into larger developments can serve to alleviate some of the difficulties found in connecting the movement patterns between the street and public interior passages. Matthijs de Boer states that "a good public interior lies on an urban route, and is therefore easily accessible" (1993, p. 19).

City streets, always important in the daily operations of the city, will remain a polyvalent space if there is a supportive, open policy to receive innovative engagement of temporary programs. The footbridge, as a transportable and rapidly installed connection device, supports the idea that a portion of a construction site or underused area may become temporarily "freed" from its operative logic for temporary events/installations (Figure 11.8). The elegant design solution of the footbridge, located mid-way between the public (the park) and the private (the Alter Hof residence) spaces, is reminiscent of Mari's description of what graffiti does to a public wall – it is the "intersection of public and private, a way of inscribing an architectural, monumental structure with the immediate concerns of an individual" (1997, p. 68). Similarly, Gordon Matta-Clark's thoughts regarding the *ultimate* transportable architecture share similar concerns: "Social mobility . . . is the greatest spatial factor . . . how one manoeuvres in the system determines what kind of space [one] works and lives in" (Lee, 2000, p. 24).

11.8 Preparing to unload the pre-fabricated footbridge

References

Barthes, R. (1998) *Camera Lucida: Reflections on Photography*. New York: Hill & Wang.

De Boer, M. (1993) "Public Interiors," in M. Kloos (ed.) *Public Interiors, Architecture and Public Life Inside Amsterdam*. Amsterdam: Architectura and Natura Press, pp. 9–32.

Fitchen, J. (1989) *Building Construction Before Mechanization*. Cambridge, Mass.: MIT Press.

Frampton, K. (2001) *Studies in Tectonic Culture*. Cambridge, Mass.: MIT Press.

Hall, E. T. (1990) *The Hidden Dimension.* Toronto: Anchor Books.

Joseph, F. and Look, J. (2003) *Tadashi Kawamata, Bridge and Archive*. Bielefeld: Kerber Verlag.

Koolhaas, R. (1994) *Delirious New York: A Retroactive Manifesto for Manhattan*. New York: The Monacelli Press.

Kostof, S. (1999) *The City Assembled. The Elements of Urban Form Through History*. New York: Bullfinch Press.

Lee, P. M. (2000) *Object to Be Destroyed: The Work of Gordon Matta-Clark*. Cambridge, Mass.: MIT Press.

LOT-EK website: http://www.lot-ek.com/permanent.html.

Mari, B. (1997) "The Art of the Intangible," in F. Bradley (ed.) *Rachel Whiteread: Shedding Life.* New York: Thames and Hudson, pp. 61–73.

Thalgott, C. (1999) "The Munich Perspective – Planning as a Process," in *The Munich Perspective: A Summary of the 1998 Urban Development Strategy*. Munich: City of Munich Department of Urban Planning, pp. 2–5.

Viso, O. M. (1996) "Charles Ray: Abstract Sculpture and Lived Reality," in N. D. Benezra (ed.) *Distemper Dissonant Themes in the Art of the 1990's*. Washington, D.C.: The Smithsonian Institution, pp. 78–85.

Wadleigh, H. R. (1910) *Munich: History, Monuments, and Art*. London: T. Fisher Unwin.

Mobility between Heaven and Earth

Meindert Versteeg

This chapter focuses on the concept design of a portable Rover/Habitat which was conceived as part of the NASA support mission to Mars. Important design aspects of this planetary workstation are the flexibility of the interior and its ability to contribute to the well-being of the four crew members. A key design ambition was to create a feeling of home in space and allow crew members to adapt their own work/live areas. When travelling on earth, we constantly find ourselves in mobile transportation environments such as trains, boats and airplanes, and are accommodated in commercial environments such as hotels, motels and inns. These terrestrial spaces along with certain key influencing influences – such as the kit-of-parts, and the theory of social logic of space – informed the design of the Rover/Habitat.

Spatial logic of the *tipi* and *yurt*

Significant examples of the first built environments which offered basic accommodation with simple construction are portable structures such as the *tipi*, *yurt* and tents. Though simple, these are very flexible and effective structures. The form of a *tipi* can evolve with changing weather conditions: by positioning the timber supports at various angles in a conical shape on the ground, it can be closed or opened, flaps can be adjusted to control wind and airflow, and a protective barrier can be installed around the periphery. Insulation can be added between the double skin and a fire can be lit in the hearth to keep the interior warm.

The interior layout of a *tipi* follows a strict spatial arrangement within a limited interior space. Allowing its inhabitants to gather around the central area where the hearth is positioned, the traditional interior layout shows designated areas for men, women and children including a separate place for male and female guests (Faegre, 1979). The Asian *yurt* uses the same spatial layout principle. Although some interior compartments change use over time (television may take the place of *kumis* bags!) the spatial arrangement is still more or less the same. Inhabitants and their belongings have designated places in both the traditional interior and the modern version. The inner space is formed around the central source of heat, situated directly under the smoke hole in the roof (Pearson, 2001). At the same time, the hole in the roof acts as a sundial and forms a sacred space inside the tent. The altar is the most sacred place in the interior and is fixed on the northern side. As in the *tipi*, the interior space of a *yurt* is built on the social and cultural differences of the inhabitants and their visitors, showing a strict separation between men and women. The structure embodies the relationship and identity of its dwellers while providing a spatial enclosure that organizes their daily life (Hillier and Hanson, 1993).

Background: kit-of-parts

The term 'kit-of-parts' describes a systematic method of building a structure from prefabricated elements (Scott Howe, 2002). During the 1920s and 1930s, new materials and techniques brought enormous changes in construction methods. The assembly of prefabricated panels and units, similar to an automobile assembly line, also influenced the construction of portable, mobile and small structures. Invented by Wally Byam (1896–1962) in the mid-1930s, a well-known example of modern mobile design is the Airstream, which incorporated a kit-of-parts prefabrication method in its construction process (Burkhart and Hunt, 2000). The first trailers were made of plywood: however, this changed rapidly when it became possible to attach an aluminium skin to a trailer chassis. The 1963 Airstream fleet introduced a wider selection of trailer sizes enabling more flexibility in interior layouts. Increased headroom was achieved by lowering the floor between the wheels. Over the years, various Airstream models showed wider circulation space achieved by spacious floor plans coupled with better climate control, resulting in a comfortable yet limited work/live space. Airstream trailers are still used by NASA to house astronauts in mobile situations.

Another kit-of-parts method can be found in the traditional Japanese house (Fawcett, 1980). This method of construction is regulated by a fixed component: the *tatami* mat – a modular building element which has a standard dimension, 900 by 1,800 mm (2.95 by 5.91 feet). This module is the key to achieving internal flexibility – interlocking *tatami* mats forming a principal component in an ancient prefabricated building system. In the early 1960s, the Japanese Metabolists offered a modern interpretation of this design philosophy. Inspired by industrialization and technology, this group of architects formed a design school in 1964 to formulate a response to the undesired effects of urban sprawl. Most of their concepts were based on prefabricated living cells that functioned as 'plug-in' living environments within a mega-structure where transformation and flexibility were the key design elements. Their English counterparts were Archigram who like the Metabolists and other architectural movements in the 1960s and early 1970s, experimented with prefabricated modular-living, box-like units and transportable environments. These environments were either self-supporting or could be connected to a larger structure, thus providing all the necessary services for their inhabitants. Japanese architects are renowned for their ability to successfully merge old building methods with new ones, exhibiting an ability to be simultaneously innovative and traditional (Franklin, 1978). A well-known example is the Nakagin Capsule Building that follows the aesthetic system employed by Kisho Kurokawa. This design is based on Japanese interlocking building components which are appropriated from their ancient wooden architecture. Each capsule in the Nakagin building is a complete living environment with a bathroom, a double bed, a desk, storage, audio and visual entertainment, and a stove, and measuring 2,400 by 2,400 by 3,800 mm (9.1 by 9.1 by 12.47 feet).

The Rover/Habitat design

These design theories and concepts informed the Rover/Habitat design which is a concept for a planetary mobile workstation that consists of two parts: a mobile rover and a fixed habitat. Joined together, the stowed-away configuration forms a cylinder that can fit into a rocket for transportation. This mobile space workstation is designed for a small crew of three to four people for short-term research on the Martian surface. A combination of modularity and mobility leads to a flexible and sensitive design solution in response to the unique environmental and human conditions involved. The gravity on Mars is 0.38 that of Earth and therefore the interior space of the habitat can resemble similar terrestrial spaces and orientation is less problematic than in zero gravity; for example, sleeping can take place on a horizontal surface. Six key factors were identified for the Rover/Habitat design that directly influence the well-being of the crew members and consequently formed the context for this design project:

- the design of the Rover follows its function
- the design and function of the Rover's interior follow its exterior interface
- the number of crew members directly affects the conditions above

- the goal of the mission affects the interior layout
- physical and psychosocial needs of crew members present a key factor for the spatial layout
- human factors, body size and social dynamics inform the design of the interior space

The simulation of a feeling of 'home' is crucial in the creation of a personalized living environment in outer space. Under normal circumstances people are engaged in a complex social matrix that links them with family members, friendship groups, organizations and society at large (Connors, 1985). In contrast, the logistical, constructional and psychological problems of living in space neglect the whole idea of 'home'. The Rover/Habitat proposes that crew members should have the opportunity to feel 'at home' during their journey or research mission.

Exterior system: landing on Mars

The Mars Rover/Habitat is a highly loaded, air-supported, semi-rigid structure. The capsule is surrounded by an over-pressurized system to keep the structural membrane in place and operate in Martian atmospheric conditions. The bottom of the structure is rigid while the top remains inflatable. As with a semi-rigid airship, this concept employs a combination of conventional and inflatable technology to allow for optimum stowage during launch and yet when deployed provides a sufficient volume for living (Figure 12.1). After landing on the Martian surface, using a braking device such as parachutes (A), the two structures are separated (A–D). The inflatable structure acts as a pressure device separating the rover from the habitat (B, C), with internal control cables that ensure a controlled deployment. Once fully inflated, the supports and wheels of both habitat and rover are deployed (D) and separation of the rover from the habitat is ensured.

The main habitat and the Rover/Habitat System are inflated simultaneously. When fully inflated, it attains a diameter of 5 metres (16.5 feet) and a length of 12 metres (39 feet). The crew enters the Rover/Habitat through a pressure port that gives access from the main habitat to the Rover, allowing them to move from one pressurized volume to another. The pressure port serves the same function as the temporary docking port that joins the station to the space shuttle (Cohen, 1996).

Interior system: a view inside

Working and living in small environments requires special design solutions. Efficient arrangement and grouping of activities is the key to making a space functional. Besides a well-laid-out floor plan, spatial comfort also promotes efficiency. Three circulation patterns are mandatory in a planetary work/live environment. These are:

- circulation paths of a crew member to/from his/her work area
- circulation paths of a crew member through the space, particularly during relaxation time
- circulation paths dedicated to the handling of liquid, solid or gaseous samples collected during a field mission for examination

An open floor plan enhances a functional and flexible interior. Allowing changes to the interior space over time is not only of psychological benefit for long-term occupation but is also essential for easy adaptation for future requirements. Design of an open and flexible floor plan involves consideration of modular functions based on the results of functional relationships and analysis.

The relationship between the work environment, living and the personal crew quarters has to be addressed appropriately. Body position is crucial in the work environment where the crew spend most of their time. The interior must support or guide the human body during work activities. In the living quarters, the interior must provide a more relaxed atmosphere, with sufficient light, wall surfaces and

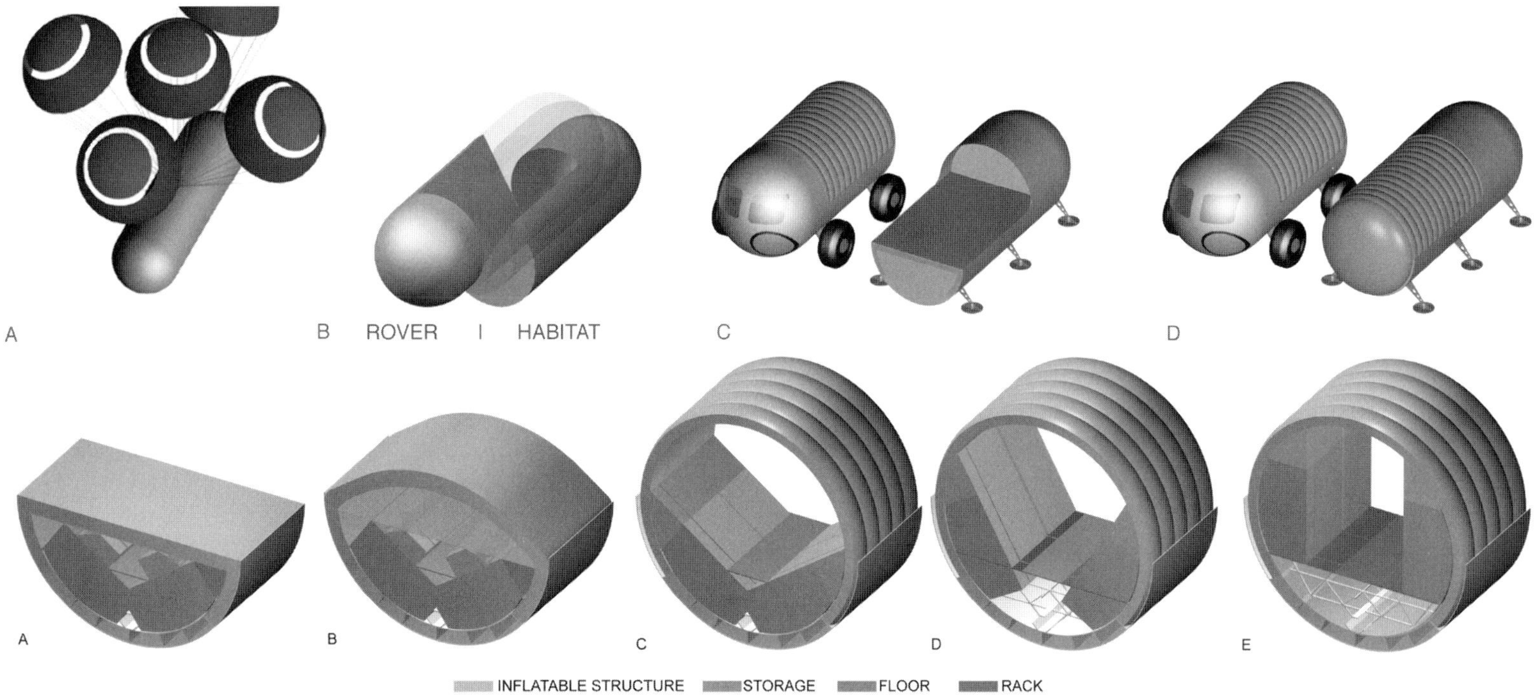

12.1 Deployment of the exterior and interior systems in the Martian atmosphere

even fragrance and music contributing to a conducive atmosphere. Lastly, the interior of the personal crew quarters, i.e. their retreat, should be adaptable to personal needs. The limited space of the Rover/Habitat's work-live environment makes it even more challenging to develop an interior interface and to evaluate feasible proposals. Important design elements include the adaptability of the interior space to suit each individual crew member's needs and the flexibility in the order of the racks of 'furnishing units' to allow for changes over time that suit the requirements of each specific mission. Rover/Habitat has to successfully accommodate circulation space, work space, living/dining space, sleeping/private space and meal preparation space. Crew satisfaction is of the utmost priority as each member needs relaxation, privacy and social contact to perform well.

In the proposal, crew members adapt the interior by positioning the foldable upper bins, located on top of the racks, and by bringing all needed elements into the racks (Figure 12.1). The lower parts of the racks are manoeuvred into position in the same way as in the interior of a car. The free space under the floor can be used for storage. Functions are grouped horizontally to provide a better work environment while also allowing separation of the working and living quarters (Figure 12.2).

The adaptation of the Rover/Habitat environment shown here demonstrates variations in space by using flexible wall elements that are positioned beside and in between the shower and washroom separating the live/work spaces from service areas (Figure 12.3). Crew members can change their own environment depending on their functional, personal or group needs. An enclosed space can be created in front of the shower to create a private entrance/dressing space or a closed habitation space that offers a larger working unit. The various spatial configurations created by the folding walls can be changed in the course of a twenty-four-hour cycle (Figure 12.4) to simulate the daily routine on earth and thus create a feeling of 'home'.

Colour can be used to delineate various functions. The use of one colour for the entire floor optically underscores the circulation zone and elongates the view despite the physical restrictions in space. In contrast to walking over the red planet, the blue floor is intended to simulate the psychological link with the blue planet Earth. In the laboratory, a sea green is used to simulate a clean and calm work environment. In the living quarters, a light hue of yellow is used to promote a relaxed atmosphere. With the upper light blue bins, the linear and horizontal arrangement elongates the length of the interior. The inner blue coating of the inflatable structure also references to the colour of the sky on Earth. It is intended that the use of different colours can help the crew members to orient themselves in a way that they are used to at home (Figure 12.5). The inner surface of the habitat can function as a 'sky', allowing a variety of lighting conditions with variable intensity. A light source that follows and projects a twenty-four-hour cycle on Earth can be a stimulus for achieving a good biological rhythm in space. A more direct illumination is chosen for the living areas to provide a more relaxed ambience when needed.

The racks, zoned according to their functions, can be positioned beside each other or on opposite sides according to ergonomic parameters. The zoning also encourages social interaction among crew members. The rack that contains the Glovebox (GB) and the rack that contains storage for solid and liquid samples (STSLS) are positioned beside each other for an easy hand-over in the two Gloveboxes inside. The racks that need water, the kitchen (KI) and shower (SH), are grouped on one side of the module to control the water circulation in one part of the Rover/Habitat system (Figure 12.6). The most dominant functions are the airlock (AL) for the crew and the positioning of the quick solid,

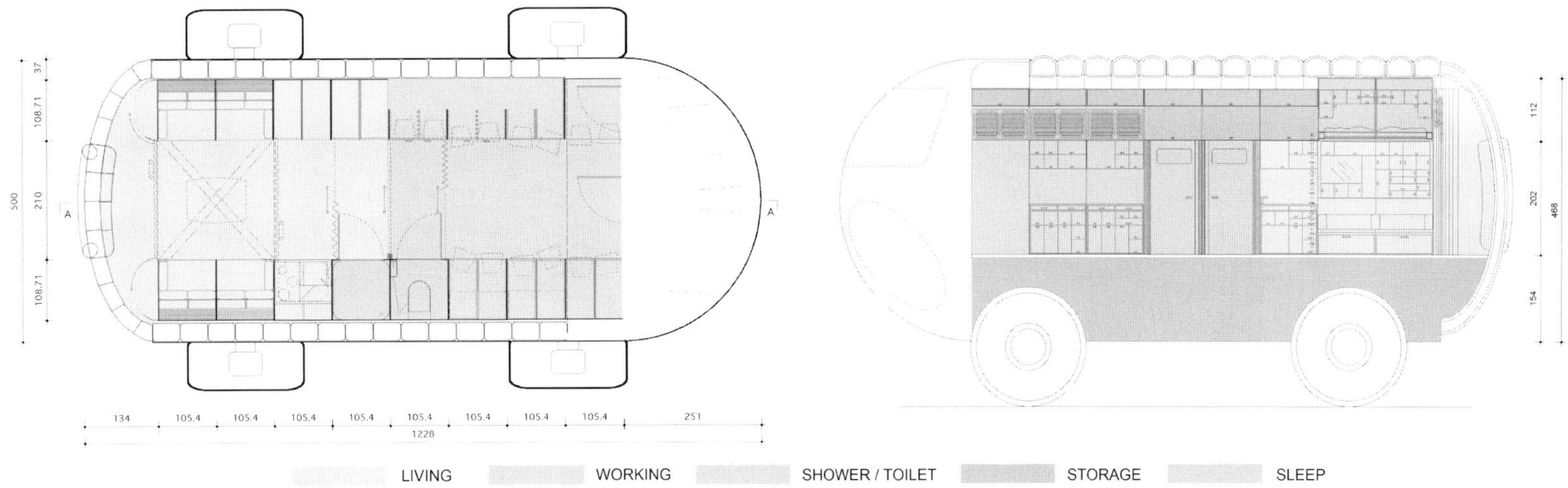

12.2 Mars Rover/Habitat system plan and cross-section

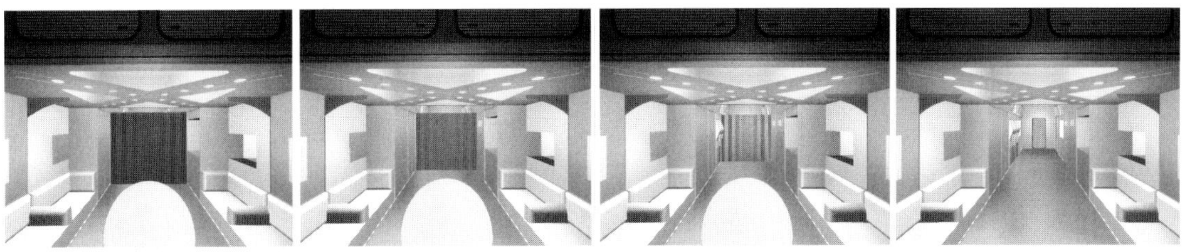

12.3 View from the living zone towards the laboratory showing the flexible wall elements

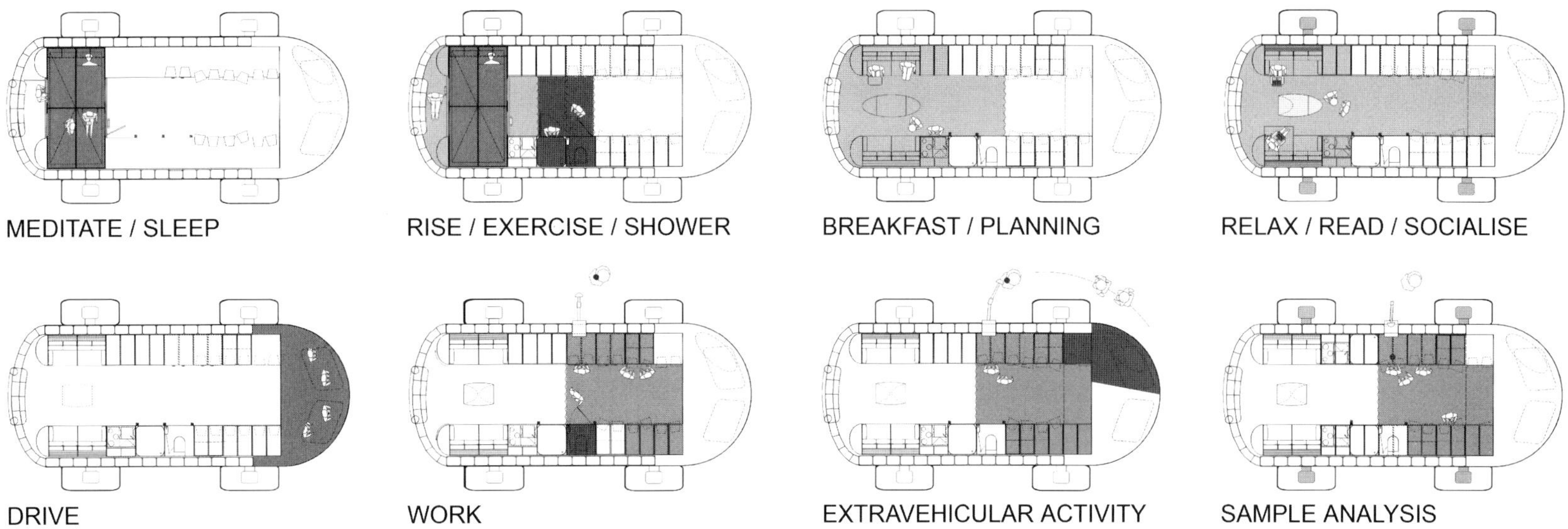

12.4 Various spatial configurations during a twenty-four-hour cycle

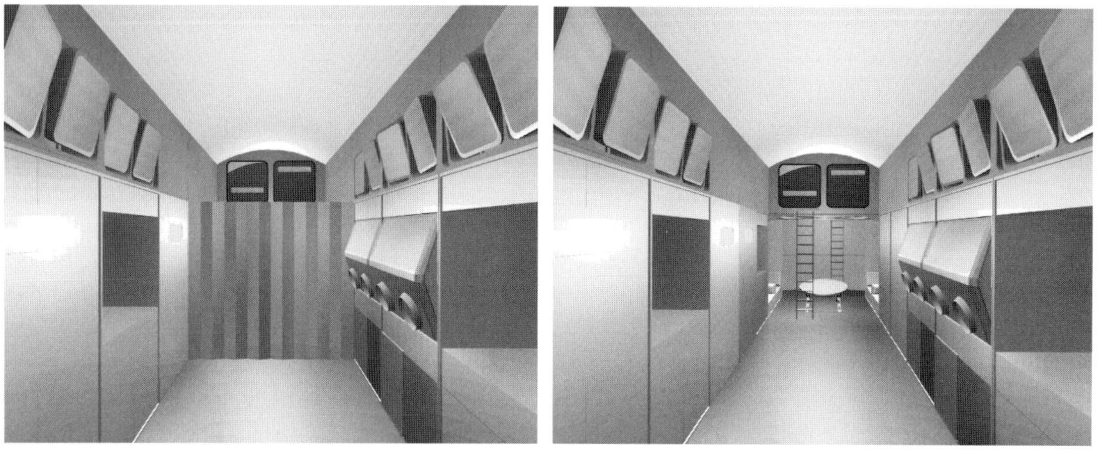

12.5 Foldable wall element separating the laboratory from the living zone

12.6 Positioning and use of the racks

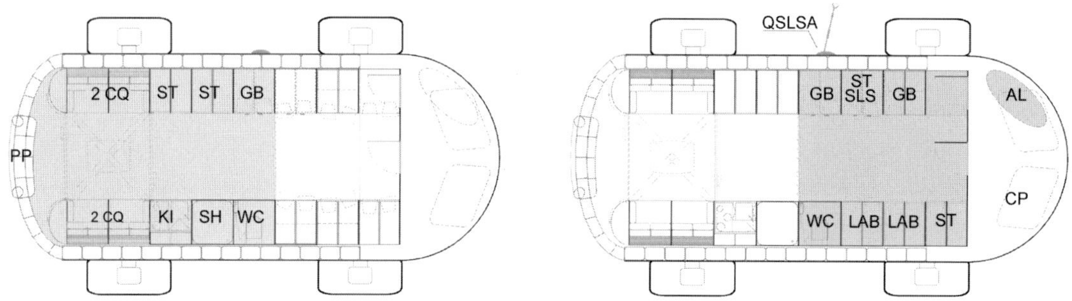

i. LIVING PATTERN

ii. WORK PATTERN

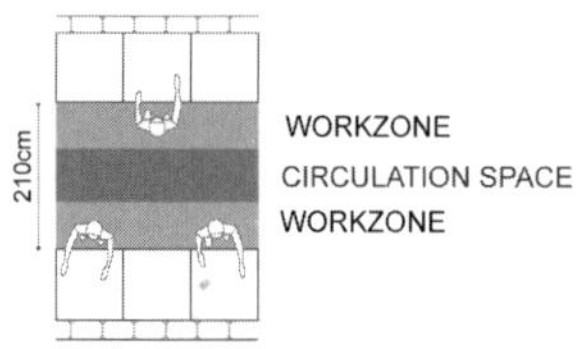

iii.

LIVING		WORKING	
CQ	= CREW QUATERS	QSLSA	= QUICK SOLID/LIGUID SAMPLE AIRLOCK
SH	= SHOWER	STSLS	= STORAGE SOLID/LIGUID SAMPLES
WC	= SANITAIR	GB	= GLOVEBOX
KI	= KITCHEN	LAB	= LABORATORY
ST	= STORAGE	CP	= COCKPIT
PP	= PRESSURE PORT	AL	= AIRLOCK

12.7 Closed living environment with illuminated ceiling (left) and door opening into private areas (middle and right)

liquid sample airlock (QSLSA) for samples that are taken from extra-vehicular activity or picked up with the robotic arm.

Human factors

Individuals from various cultural backgrounds with different physical requirements and ages participate in space programmes. This makes gathering relevant anthropometric data prior to the design process very important in order to design objects, surroundings and interfaces to fit their specific human skills, limits, behaviour and motivation (Messerschmidt and Bertrand, 1999). Accommodating the diversity in body measurements is very crucial for safety and comfort as, in extreme situations, an 'unfitted' interface between human and machine can cause life-threatening circumstances. The Rover/Habitat concept therefore accommodates an appropriate range of required ergonomic dimensions that are 'fit' for male and female and large and small crew members.

Spatial division of the living area

An adaptable cooking space can cater to the different needs of the crew and sustain them over a longer period of time. The cooking space is situated beside the shower to optimize water circulation, thus avoiding long noisy runs. A working surface with a small sink can be pulled out and turned away for easy access to the cooking rack. Another rack with a workplace and basic storage is positioned on the opposite side. The dining area, while providing space for crew members to have their meals, also provides a place to socialize as well as a temporary additional workplace when needed. The space, situated under the private crew quarters, generates a more enclosed environment. The inclusion of an elliptical and multifunctional table allows easy movement and access. Light intensity can either vary to simulate the time of day on Earth or provide 'a personal sky' (Figure 12.7).

While allowing a crew member to retreat from others, the personalization of sleep and private areas becomes essential in making spaces comfortable during a journey into space. For crew members, decorating their space with personal belongings, photos or other personal items can evoke a connection to their home, family or relatives. Further visible variations in texture, colour and materials can contribute to the well-being of the individual. Access to each of the crew quarters is provided via a ladder. When open, a large door provides a physically larger space without sacrificing too much from their privacy (Figure 12.7). While small personal items can be stored inside the private areas, clothes and other larger items are stored in the storage volumes at the end of the living and dining spaces below.

Conclusion

Designing enclosed environments such as small space capsules is a challenge due to the restrictions imposed by the limited volume. Achieving a sense of 'home' must be an integral part of the design strategy. By using flexible elements, foldable walls, colours and special lighting features, crew members can adapt their live/work environment to their needs. In this way, stress caused by living in strange surroundings and feelings of isolation can be minimized. It is an important task of the architectural and environmental designer to shape and reshape the surroundings in the space capsule to meet the psychological, social, cultural and aesthetic needs of the crew members while also satisfying their vital physiological needs (NASA 1977).

Part of this chapter is based with permission on the SAE Paper 2003–01–2391 2003 SAE International.

References

Burkhart, B. and Hunt, D. (2000) *Airstream: The History of the Land Yacht*. San Francisco: Chronicle Books.

Cohen, M. (1996) 'Habitat Distinctions: Planetary Versus Interplanetary Architecture', in American Institute of Aeronautics and Astronautics Space Programs and Technologies Conference, Huntsville, AL, AIAA-96–4467, p. 5.

Connors, M. M. (1985) 'Living Aloft: Human Requirements for Extended Spaceflight'. Retrieved January 2001, from http://www.hq.nasa.gov/office/pao/History/SP-483/contents.html, p. 4.

Faegre, T. (1979) *Tents: Architecture of the Nomads*. New York: Anchor Books.

Fawcett, C. (1980) *The New Japanese House: Ritual and Anti-Ritual, Patterns of Dwelling*. New York: Harper & Row.

Franklin, R. M. (1978) *Beyond Metabolism: The New Japanese Architecture*. New York: Architectural Record Books.

Hillier, B. and Hanson, J. (1993) *The Social Logic of Space*. Cambridge: Cambridge University Press.

Messerschmidt, E. and Bertrand, R. (1999) *Space Station, Systems and Utilization*. Munich: Springer.

NASA (1977) *Space Settlements* 'Architectural Studies for a Space Habitat'. Princeton University Conference Meeting no. 127, on Space Manufacturing Facilities, Princeton University. Special issue, electronic version retrieved April 2003, from http://lifesci3.arc.nasa.gov/SpaceSettlement/75SummerStudy/Table_of_Contents1.html.

NASA (1998) 'Suited for Spacewalking: A Teacher's Guide with Activities for Technology Education, Mathematics, and Science', retrieved November 1999, from http://spacelink.nasa.gov.

NASA (2004) *NASA-STD-3000*, Volume 1, Section 8. Special issue, electronic version retrieved June 2004, from http://msis.jsc.nasa.gov/sections/section08.htm#_8.2_overall_architectural.

Pearson, D. (2001) *Yurts, Tipis and Benders: The House that Jack Built.* London: Gaia Books.

Scott Howe, A. (2002) 'The Ultimate Construction Toy – Applying Kit-of-Parts Theory to Habitat and Vehicle Design', in American Institute of Aeronautics and Astronautics Aerospace Architecture Symposium, 6116. Houston, TX.

Versteeg, M. E. and Veldman, S. L. (2003) 'Transporting Space: A Concept Design for a Planetary Mobile Workstation', in *SAE International,* Engineering Society for Advancing Mobility Land Sea Air and Space. 2003–01–2391. Vancouver, AB.

Mobile Architecture and Pre-manufactured Buildings: Two Case Studies

Dean Goodman

Introduction

The size of the homeless population in the city of Toronto has grown to an estimated 4,000 people on any given day. This figure represents the people currently moving between emergency shelters and the street. The political will for taking action to provide stable, relevant housing for this population has been and continues to be inadequate. Furthermore, the housing that has been built has, for the most part, relied on a formulaic architectural response – one that is neither sensitive nor appropriate to the particular needs of many in this group of individuals. In addition, the procurement method, from assembling land through to the construction methods employed, is not flexible enough to tackle the magnitude of this ever-growing problem. Consequently, the extent of homelessness in Toronto has given rise to the necessity for architects to investigate and begin to adopt alternative means to construct housing in a quick and flexible fashion.

This chapter will investigate two case studies which have attempted to provide housing for the homeless population in downtown Toronto, using mobile architecture and pre-manufactured housing. Both are proposed because of the rapidity of construction, flexibility in design and sophisticated construction technology used in their manufacture. The first example proposes the use of mobile architecture to build a temporary community at an abandoned brownfield site along the waterfront. The second involves pre-manufactured units to be built on the roof of an existing four-storey, not-for-profit apartment building. Both are examples which attempt to break the mould in mobile architecture and pre-manufactured buildings. Neither involves unique or custom-designed modular buildings or expensive production-line architecture such as travel trailers. In addition, they are not an off-shoot of the 'single family' residential model currently associated with standard mobile homes. Both are, however, urban; located in the downtown core of the city of Toronto; specific to the cost benchmarks of non-profit residential housing; relatively low cost; and strategically and specifically designed. Both projects address the issue of the applicability of modular buildings in urban settings. They challenge the assumptions about the role of design for both the architect and consumer of the 'product' in our society. Finally, the chapter examines the ingrained prejudice in the building industry against the proposition of employing modular technology for this application. This is evident in the employment of 'high design' in such examples of mobile architecture as travel trailers, boats and even custom modular buildings as compared to the obvious lack of design consideration – except in service of strict functional criteria – in mobile products such as house trailers, portables and even port-a-potties.

The trailer and trailer parks

Mobile architecture has a rich history in western culture but is predominantly associated with two types embedded in popular imagery of postwar society. The recreational application in the form of travel trailers or caravans, recreational vehicles (RVs) and boats is one common example. A particularly relevant off-shoot of this is house trailers which, although they are a form of mobile residential architecture, once based in a trailer park the mobile unit becomes part of a permanent or semi-permanent community. The other commonly known area of mobile or modular design involves the use of modular buildings, for example as school classrooms or construction site facilities. These latter are typically conceived of as semi-permanent occupations.

One of the most compelling aspects of mobile architecture and pre-manufactured buildings is that they are so prevalent and yet so faceless. Most cities and towns in North America have significant trailer park communities. Though these are relatively permanent (they have underground services, are taxed as property, have amenity and recreation services such as community centres), they are perceived by society at large as poor cousins to 'real' housing. The picture is quite different in rapidly growing or new communities where the rate of growth outpaces the availability of the existing infrastructure required to construct 'authentic' suburban housing. An example can be seen by examining what occurred in Prince George, British Columbia, during the 1970s. This resource-based city in the northern part of the province experienced phenomenal urban expansion. The growth was so rapid that a large sign at the city limits boasted that more than half the area's population lived in house trailers. The municipality was very proud of their city's new viability and desirability, so much so that it publicly celebrated the new trailer park communities as a sign of wealth and potential – a far cry from the contemporary view so well expressed in recent television shows such as *Trailer Park Boys*.

Both for the semi-permanent modular buildings and portables that are located in many school yards and construction sites and for house trailers, architects have either been by-passed in the design stage or have tended to shy away from embracing their inherent design opportunities. A parallel situation can be seen in the general attitude the architectural profession has towards the design of suburban housing. For numerous reasons, such as the design of the housing being subservient to the marketing imperatives determined long before architects gets involved; by the professions' self-editing as to authentic versus derivative design; or the economic infrastructure of suburban developments that predetermines much of the traditional input of the architectural profession, the profession has mostly failed to make design inputs relevant in these buildings.

From a design perspective, architects by and large are dismissive of the potential inherent in house trailers and the urban design potential of mobile home communities in general. People who attend school in portables, work at construction sites and/or live in mobile homes are not seen as consumers of real design. On the other extreme is the mobile architecture of travel trailers, RVs and boats. In this branch of the typology, design does play a significant role. There are a plethora of sophisticated designs and each is both planned with innovation and built with a variety of levels of craftsmanship. They have become classics in modern culture. Clubs are formed and members plan trips, often being hosted by affiliate club members in whatever jurisdiction they are travelling in, to places where a caravan of trailers and RVs travel significantly long distances. These clubs are international in breadth and have as much popularity in Great Britain and Europe as they do in North America. In New Mexico, there is the Shady Dell, a motel composed of a vintage collection of classic Airstream trailers. The corollary to this area of mobile architecture, where design issues are paramount, is custom-built pre-manufactured houses. These are unique one-off houses where extreme care is taken in terms of both design and construction techniques.

Technology

The common link between these examples is that both mobile architecture and pre-manufactured building are both constructed in factories. The technological advantages are numerous and if a proper understanding of and appreciation for this type was developed in the architecture community, the potential could provoke further development. The existing knowledge in the industry that allows the manufacturer to achieve a high level of technical sophistication and quality control is worthy of further review.

With respect to this chapter, it is of interest to highlight the way in which this type of process can be used by architects and builders to provide options for built form and/or a superior end product for the same construction value. Construction that occurs in a factory is not subject to the vagaries of weather, site-specific conditions or the high turnover of construction workers. Proper construction techniques are more likely to be ensured when one is working inside a factory that is well lit and temperature controlled and where the work is done standing on a concrete floor rather than off scaffolding six floors above ground level, in the middle of winter. Factory conditions allow for a more consistent end product, one that typically only occurs with a great deal of effort when the project is constructed on site and has a high end construction value rather than if it is not-for-profit construction which has values typically at the low end of the spectrum.

Mobile architecture: Tent City (Levitt Goodman Architects, Homes First Society, Jon Harstone Housing Consultant, 2002)

Tent City was the name given to a squatters' community that existed in Toronto between the years 1997 and 2002. The location was a brownfield site in an abandoned industrial district at the south-east end of the city, where the Don River meets Lake Ontario. A survey taken in the autumn of 2001 showed that there were more than 175 residents living full time at Tent City, with the population swelling in summer and shrinking to the core community in the winter months (Figure 13.1). As the nucleus population became more settled, a next generation of squatters sought out the relative stability and jurisdictional freedom of this squatters' camp. At the time of the 2001 survey, there were a variety of house forms including plywood shelters, tents, camper trailers and wooden frame constructions from scrap dimensioned lumber that was wrapped in clear plastic. The size of the houses ranged from 9 to 31 square metres (100 to 350 square feet). An infrastructure had been informally developed of main and secondary streets, front yards, a bath house and a centralized 'renegade' generator that provided electricity to the community for a number of hours a day. However, the land was highly polluted and had no running water or amenities that could sustain a healthy existence.

In various consultant reports and articles in the media, the people who lived in Tent City were referred to as homeless; however, if one consulted the core population of Tent City they did not refer to themselves in this way. Our firm, Levitt Goodman Architects, was involved in the analysis of the community, the infrastructure and the governing structures that had been implemented by the residents. We determined that many of the residents were better housed at Tent City than at the majority of shelters that would be their only alternative should they have availed themselves of the conventional residential options. Uncovered was the underlying tension between the *ad hoc* occupation of land by a disenfranchised group of citizens who are not normally accorded the right of self-determination and the more politically acceptable, i.e. stable, *tax-paying*, citizenry who viewed Tent City as either an abomination or an aberration or both. Referring to the residents of Tent City as homeless denied them the legitimacy of community. In addition, during the late 1990s the issue of homelessness was becoming more visible in general as the number of homeless adults and families grew exponentially. Tent City became the poster image in a politically charged

13.1 Original house at Tent City

13.2 View towards Lake Ontario

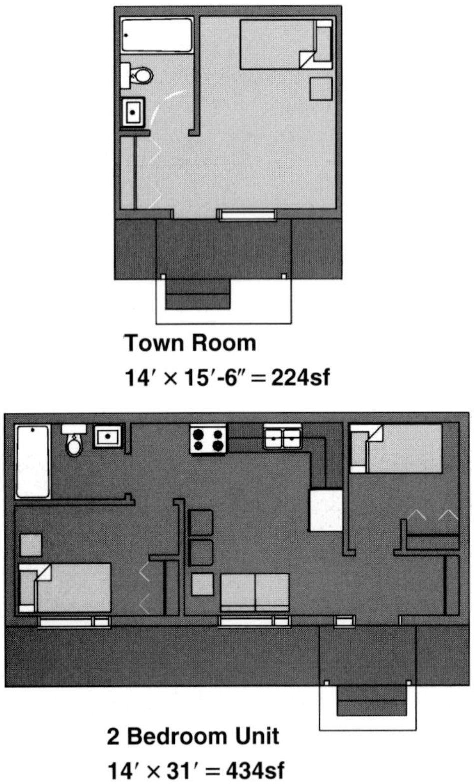

13.3 Proposed unit plans

environment where people were calling for action with highly diverse solutions. What both ends of the political spectrum were in agreement about, however, was that this was a problem that the government – Federal, Provincial and Municipal – needed to solve.

In 2001, a culmination of political and public outrage galvanized the city into taking action. This was in part fuelled by the fear of the municipality that residents of Tent City might begin to demand utilities and hold the city and the land owner jointly responsible for cleaning up the polluted site for their use. To begin to address the problem, the city tendered a proposal call in the autumn of 2001 to build housing for this community. The life span of the built form was to be three years, the time frame deemed necessary for the city to provide interim housing while it located a site for permanent housing. Our proposal consisted of a series of mobile buildings which, when combined together, formed three small exterior courtyards facing Lake Ontario (Figure 13.2). In view of the lack of site services available in this area of the city (as it was an abandoned port neighbourhood), we proposed onsite storage tanks for sewage which would be pumped out by trucks on a regular basis. We addressed the need for potable water by using water trucks that would pump water into temporary cisterns. These solutions alleviated the need to install underground services to sustain the community for a three-year time frame. The proposed individual houses were approximately 21 square metres (224 square feet) each and a series of four units made up one modular building. Front porches running the length of each building were designed to provide a semi-private threshold between the public courtyards and private apartments. Each house would be self-contained with a bathroom and kitchenette. In order to accommodate couples and families we incorporated an interior door that was designed to link two or more of the units together, with the maximum being four. In this way the basic modular unit could accommodate various size and configurations of households (Figure 13.3).

Due to the short-term nature of the tenure at Tent City and the self-contained aspect of services, we determined that the mobile home model was particularly appropriate for the method of design and construction of these buildings. We also projected a scheme for the recycling of these units after the three-year time frame had expired. We proposed reusing these units in another context that could be either temporary or permanent as the infrastructure dictated. This proposition proved to be too radical to implement and various tensions came to play on the fate of the project. On the one hand, by acknowledging that we were housing people in house trailers the sensibilities of a number of civic people were deeply offended. This was perceived as only 'one step up' from the makeshift type of structures currently on the site. There was also the concern many voiced that by making the buildings permanent and employing a pre-manufactured construction, the wrong message was being sent to other under-housed or squatter communities watching this situation unfold. Those who voiced this type of reservation about the project were of the opinion that it was politically unacceptable to give any land to squatters, especially waterfront property. This was regardless of the fact that the proposal was temporary and meant to ensure safe housing on an interim basis for those living at Tent City.

Finally, the loss of nerve at the municipal government level ensured the demise of Tent City. Before our team had finished the design stage, a series of obstructions were put in its way. For instance, it became impossible to ascertain the criteria upon which the city would release a building permit, or to obtain a consistent reading of the zoning which would be applied to the proposition. In the autumn of 2002, the owners of the land erected a fence around the site, bulldozed the buildings and paved over the entire site with asphalt. The residents of Tent City were offered assistance to secure shelter in one of the city facilities or if possible in subsidized housing. Some opted to return to the streets rather than live in shelters.

Pre-manufactured housing: St Claire's Multifaith Housing Society (Levitt Goodman Architects, Jon Harstone Housing Consultant, 2004)

In 2000, St Claire's Multifaith Housing Society purchased a four-storey, forty-seven-year-old medical building at 25 Leonard Avenue in the Kensington Market, a vibrant neighbourhood in downtown Toronto (Figure 13.4). We were the architects involved in its conversion into a fifty-one-unit apartment building. The tenants are single adults who had formerly been living in various city shelters. For many, the apartments they moved into on Leonard Avenue were the first private accommodation they had had.

In 2003, St Claire's decided to expand their housing programme and began to look for additional land in the same neighbourhood. They soon discovered that the primary impediment in developing and building new non-profit housing in the city is the availability of affordable land in a suitable neighbourhood. Two preconditions for securing funding from the city agency responsible for the allocation of grants reinforce this predicament – the requirements that housing providers have land available or an accepted offer, and that the housing group is able to meet specific timetables for the delivery of completed units prior to final approval of funding.

During the conversion of the building, we discovered that the existing structure had been designed to allow for an additional number of floors at some future point in time. The idea of maximizing the existing site by adding new apartments on the roof of the existing building, using pre-manufactured units resulted from the housing group's obligations to meet the city's funding requirements (Figure 13.5). Due to the pre-existing building condition that St Claire's owned 'developable' land and the rapidity with which pre-manufactured housing is constructed, they were able to meet delivery targets. Furthermore, the programme addressed additional concerns of the team – to ensure that the new construction was of excellent quality and could be built during winter on the fifth and sixth floors, and to minimize disruption to the existing fifty-one tenants who would continue to inhabit the building during construction.

Meeting these concerns has led us to look at this project as a case study prototype. We have been looking for different situations which will allow other housing providers in the city who already own buildings to intensify their sites through the application of pre-manufactured housing principles. Many housing organizations can finance the pre-development costs necessary for rezoning a site to increase density if the land is not sitting vacant. In addition, if and when the project receives funding, the time that is required to assemble a set of construction documents for a pre-manufactured housing assembly is approximately half the time that is required for a conventional permit set. Finally, as the pre-manufactured buildings are built from steel studs, this provides fewer design challenges and proves to be less costly regarding the additional loads it imposes on the existing buildings' structural system. In addition, the inherent flexibility of the pre-manufactured module allows a better fit within an existing structural grid with a minimum of upgrades.

At 25 Leonard, our design approach was to organize the new apartments around an exterior landscaped courtyard located on the roof of the existing building. The 'garden apartment' typology was examined as a design precedent regardless of the fact that these were to be located on the fifth and sixth floors of the building (Figure 13.6). We planned to extend the existing elevator core as well as both

13.4 Original building on 25 Leonard Avenue

13.5 Elevation with garden apartments above

13.6 View of courtyard, fifth floor

exit stair towers to service the new floors. These new service cores divided the main exterior landscaped space into three smaller courtyards with the apartments flanking the courtyard in the form of two rectangular bars of units running the length of the building in an east/west orientation. The size of each of these bars was determined by the optimum size of pre-manufactured housing unit that could be transported to the site by a truck. The individual apartments were accessed from the two floors of covered exterior walkways. This arrangement allowed for excellent natural light and ventilation to each unit as well as ensuring equal access to the shared walkways and the common garden courtyard. As air conditioning is not funded in non-profit housing, the exterior walkways serve as balconies for each unit, providing ample exterior space during the hot summer months and a source of cross-ventilation for each unit. Each apartment was based on single room occupancy (SRO) and 19.5 square metres (212 square feet). Including the exterior walkways to the apartment increases each unit's available area to 26.5 square metres (285 square feet). Each unit including the walkway measures 3.5 by 7.5 metres (11.5 by 24.75 feet) and was to be built and shipped individually from the factory to the site. After their arrival at the site, a crane would lift each unit directly off the truck and stack them like large-scale building blocks in their predetermined location on either the fifth or sixth floor. The rate of assembly was planned to be one apartment per day.

The modular units arrive completely finished including appliances, light fixtures, millwork and a built-in kitchen table. We studied a range of Airstream travel trailer designs to develop strategies for minutely 'programming' a small space that maximizes storage while still creating a generously proportioned interior. Consequently, we have designed many more built-in furnishings than would ever be included in a not-for-profit unit. Appliances and fixtures are tucked into alcoves and corners rather than being located in separate rooms. The bathroom is not a discrete, stand-alone room but is broken down into its constituent parts. There is no dedicated bathroom sink in the design, rather the resident uses the kitchen sink for all washing purposes. The bathtub has glass sliding doors creating its own enclosure that opens directly into the main space. The kitchen is small but incorporates full-size appliances and the built-in kitchen table does double duty as kitchen counter as well. Outside each unit's front door along the walkway is a built-in bench, offering residents a place to sit, to store gardening materials and/or to put down packages while opening the front door to the unit. Given that the tenants moving into these apartments are arriving directly from a city shelter, they have few possessions with which to set up a house. Storage space is always scarce and in small units this is especially so. In this example, we have made the ceiling height 2.5 metres (8.5 feet) which allows us to build in storage along the entire length of a 'service wall' including over the entry and bathroom fixtures. Though there is still room to personalize the space, the interior of these units is designed to be finished to a higher degree than typical apartments.

Conclusion

Tent City and St Claire's are practical projects which demonstrate the adaptability of various forms of pre-manufactured buildings to address the issue of housing shortage in downtown Toronto. They are prototypical applications in urban contexts with high design expectations. Although both examples were developed for the not-for-profit sector, they demonstrate the design opportunities inherent in this type of housing and emphasize applications outside this particular housing sector. In both instances, the mobile housing applied to Tent City and the pre-manufactured units for the St Claire project are unconventional construction methods that offer opportunities that conventional building methods do not.

Whom design is meant for in our society, who consumes it and how the construction industry tailors its products to serve those consumers is the underlying subtext of this chapter. Typically, as I have illustrated in the two examples, these industries cater their production to specific groups of people and it is clear where their efforts are directed. It is also clear that, in general, the architectural profession as a whole applies a similar hierarchy to its effort. In these two case studies, the fact that mobile buildings and pre-manufactured housing with a large design component are intended for homeless or under-housed people indicates the potential available if designers are able to shift their priorities. Being provided with buildings with an emphasis on design is not the typical situation for the homeless community. People on the fringes of society as well as those living in trailer parks are not traditionally embraced when discussing the merit or impetus for introducing design or technological innovation. The mobile home and manufactured housing industry are adaptable and have the inherent flexibility to meet varied design constraints.

Project credits

Architect Team: Levitt Goodman Architects Ltd. Dean Goodman (partner-in-charge), Danny Bartman, Marko Lavrisa, Alex Tedesco, David Warne
Housing Consultant: Jon Harstone
Structural: Balke Engineering Inc.
Mechanical and Electrical: Keen Engineering Ltd
Fire and Life Safety: Arencon Inc.
Construction Management: Dineen Construction

Technology

Transportable Environments: Technological Innovation and the Response to Change

Robert Kronenburg
University of Liverpool

14.1 Information overload in Tokyo's Shibuya district

Manufactured portable structures have been in existence since humankind first began to build and though their design has been shaped by many cultural, environmental and economic factors, this architecture is most cogently defined in terms of its structure and materiality. The most advanced contemporary descendants of these early, but by no means primitive, mobile structures follow similar structural principles. However, the materials that they can now be built from are very different. Membranes may be made from molecularly designed plastics, frames from complex alloys, and temporary load-bearing structures may be created from a wide range of reusable and recyclable resources. In addition, new materials are now being designed to perform different tasks besides support and enclosure – they can contribute to the building performance, operation, appearance and deployment.

Technological innovation is just one of the ways in which pressure for change in architectural design is being applied. Contemporary design must also respond to the demands of ecological and sustainable criteria, increasing legislative control, economic pressures and a perceived need for flexibility and measurement of success over time. Transportable built environments often fulfil the same function as static ones though with the added problems that inevitably result from temporary sites and servicing. The design of these facilities, with their requirement to meet these difficult performance and operational criteria, is therefore often at the vanguard of innovation in the building industry. This chapter suggests ways in which the experience gained can be measured and examines its relevance to the world of general architectural design.

I

The physical forms that shape the urban environment are produced by largely static elements – landscape, infrastructure and buildings. The common perception of this environment is that it is basically stable, largely unchanging and overwhelmingly continuous. But this is untrue. Important and profound changes continually take place in both the way that people engage with the environment that they inhabit and, as a result of their demands, the physical form it takes. Human beings are moving, changing, acting and reacting creatures that demand adaptability in everything they do. Even difficult to change things must be changeable – houses altered, highways rebuilt, skyscrapers demolished and replaced. Civilisation has an in-built proclivity for change and though this has always been a condition of human existence, it would seem that in contemporary life ever-increasing levels of complexity in the economy, society and technology have accentuated this natural condition.

It is natural in this whirl of innovation and movement to seek stability, and the relative semi-permanence of architecture has been a general expression of cultural 'permanence'. Buildings can be designed so that they express the solidity of a society that perceives it is here to stay. However, though the stone and concrete of classical antiquity are more durable, they are no more *permanent* than the tents and fabric of the nomads. Change is an inevitable development in the continuation of all things, especially those that are human-made. In the first 140,000 years or so of building design, flexibility and portability were perceived as the inevitable manner of human existence – a hunter/gatherer nomadic lifestyle was necessary for survival. In the last 10,000 years as crop cultivation and animal husbandry became established, a more settled lifestyle was able to develop which made practical the creation of more substantial and less portable buildings. Since the beginning of the industrial age, the value of transportable buildings has been repeatedly exploited and though the first human-made architecture was relatively simple domestic environments, this type of structure has since been adapted to every sphere of activity – commercial, industrial, entertainment, medicine, education and military. In the information age, interest in transportable environments has intensified. For many in the West the question 'how do we live our life?' has for the first time become separated from the necessity to be in a certain place continuously for work. Work, and meetings with co-workers, can now be done remotely. This leads to the possibility of people not only dwelling in different places, but also dwelling in movable places. In addition, services that could once be provided in a fixed location in order to operate must now be relocated to different places to perform their functions. For example, the bank, once a solid building in the centre of each town, currently a multi-national facility with scattered headquarters and branch offices, is now rapidly being superseded by mobile facilities and Internet banking. After the 1993 San Francisco earthquake the banks were the first business facilities up and running by importing the many mobile facilities they keep for expos, conventions and special events. However, our expectations of building performance have not been compromised simply because buildings may now be transportable. Contemporary mobile architecture must not only be feasible but also practical, efficient and economically viable. Regardless of these rigorous demands, there is now a wide range of mobile buildings and facilities that fulfil every task that static architecture is asked to do, and more than at any time previously people are looking to portable solutions as a viable alternative approach to solving building problem requirements.

Technological innovation has been an important driving force in the development of transportable environment design. As has already been stated, it is a cause for its renewed importance and relevance, but it is also a factor in its more ready acceptance by clients and users. Contemporary society has an interest in innovation partly from the recognition that life, on the whole, has been made easier and safer by technological advance. There is significant negative impact associated with technological advance, however, as Martin Heidegger states: 'what is dangerous is not technology – technology is not demonic. . . . The actual threat has already affected man in his essence' (Krell, 1993, p. 333). It is also a major cultural influence in terms of communication, travel and entertainment. Technology affects us directly and indirectly in many ways and is integrated into our daily life, primarily through the powerful forces of information technology and personal transport, to an extent it would have been difficult to predict even fifty years ago. The competition for our investment in these areas – personal computers; mobile phones; video and audio; computer gaming; cars; holidays – is fierce and all pervasive, only matched by the public's seemingly endless desire for them. But technology has also been crucial in the capacity for portable building to be able to respond to increased performance requirements. Like conventional building, design precedent is important, but in portable architecture, ingenuity and innovation are critical.

II

Technology has an important part to play in all high-performance building; however, the demands that result from the requirement to make a building move are extreme. This is because the mobile building must perform all the roles of a conventional one but

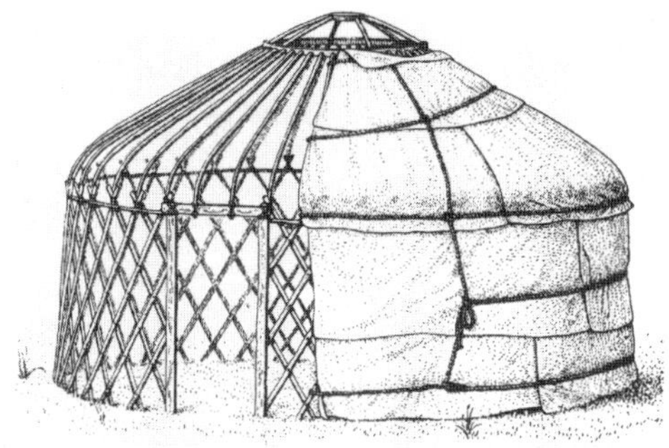

14.2 Sophistication in traditional structures: the *yurta* utilises geodesics, tension and compression rings in its structure and is manufactured using production-line strategies for efficiency

14.3 Mobile bank design for the Trustee Savings Bank by Lorenzo Appicella

in a form that is sufficiently strong and light to be relocated without excessive effort, employing sophisticated lightweight structural and constructional forms that balance deployability with usability. Balanced sophistication is a characteristic of mobile building that has persisted throughout its history: tents utilise tensile fabric engineering with minimal, relatively heavy-compression elements; tipis are a single cell of a space frame that incorporate a twin-skin environmental wall for ventilation and insulation; the *yurta* (or *ger*) is manufactured using modular systems incorporating a geodetic wall structure. Though these traditional building types (and many others) use low-tech materials, the systems they employ have been refined over generations into highly efficient sophisticated strategies. This is architecture expressive of the cultural, environmental and economic factors which have shaped it – but it also incorporates physical forms based on timeless structural principles that are still present in the most advanced contemporary architecture.

Modern technological approaches to mobile building design make use of these and other precedents – but innovation is still an essential component of making buildings do more, and do it better. Buckminster Fuller's stirring evocation made in 1946 to utilise technology to 'make "house" do much more than a house ever did before' is still an axiom that lies at the heart of the relationship between architecture and innovation (Mellor, 1970, p. 169). Transportable architecture often sets out to solve difficult problems in new ways. This is sometimes because a more efficient or economic method of achieving the objective is required – but it is also frequently because the problem simply cannot be addressed without a completely new approach. Without innovation much transportable architecture simply would not be possible, so new design approaches are a critical driving force in its successful construction and operation.

Because of the essential pioneering component in portable architecture design, it is often at the vanguard of innovation in the building industry. Commissioning is frequently done with unconventional contracts, design carried out in direct collaboration with manufacturers and builders, construction undertaken in parallel with integrated testing and commissioning procedures. These factors encourage the use of new procedures and strategies whilst at the same time limiting the possibility of failure. Plastic membranes, first used in mobile tension structures, are now regularly used in more permanent situations. Stiff lightweight compression and bending elements using complex alloys and special casting techniques were developed for mobile situations in other industries before being transferred to architecture. Materials that multi-task, providing light, power and communication mechanisms as well as environmental enclosure and shelter, have been used in mobile exhibition and entertainment structures but are now finding their way into more conventional building projects. The infiltration of new technologies into architecture is manifest. However, its effects must be measured so that its value can be assessed. Though there are numerous methods of assessing architecture through techniques such as post-occupation analysis and life-value analysis, investigation of the specific value of innovation is not usual. Study of the range of built portable architecture has led to the conclusion that there are three areas of particular significance that should be examined: application, operation, and human response. (This chapter builds on the author's earlier work, which is based on the examination of a range of case studies. The themes identified were technology transfer, logistical procedures and alternative human response (Kronenburg, 2003, pp. 7–18).)

In order for technology to be useful it must be applied. It is in this area that many problems arise because the promise of beneficial technology cannot be delivered in a sufficiently economic manner or there are issues associated with unwanted collateral effects (for example, the safety issues and environmental pollution concerns of atomic power stations).

14.4 The AT&T Global Olympic Pavilion design by FTL Design Engineering for the 1996 Olympic Games in Atlanta incorporated innovative tensile membrane modelling and manufacture coupled with special projection technology to create a building skin that was also a communication and entertainment device. FTL has since designed buildings that incorporate interactive membranes that are both shading devices and solar power generators

14.5 Festo KG: blue sky research leads to innovative buildings such as the pneumatic Airtecture hall first erected in Esslingen, Germany in 2000

14.6 Branson Coates's Powerhouse::UK – a mobile showcase for British design (1998)

Most innovative technological development in architecture results from technology transfer – the application of materials and techniques initially developed for use in other fields. In these cases any associated problems have usually been established, understood and dealt with in the initial research or subsequent primary application roles. Transportable architecture has frequently utilised technology transfer as a route to solving problems. Not surprisingly, a typical resource has been the design and manufacture of vehicles, for example the use of yacht structures in the work of Richard Horden (Horden, 1995). Only very rarely is an architectural project utilised to explore completely new technological potential – in effect to make 'blue sky' research. However, the portable building projects by Festo KG are created with just this intention as an interactive experimental process involving design, commissioning and manufacturing. Festo is an international company specialising in hydraulic and pneumatic machines used in high-tech robotic industrial manufacture. Their corporate design division has the brief of creating new products that utilise the expertise of the company and transcend its current market boundaries. These products are both realistic and useful *and* experimental and unusual. They serve as potential pathways into the future that the company might explore but also as advertisements for its present-day technological creativity (Kronenburg, 2003, pp. 71–84).

One significant way in which transportable buildings differ from conventional ones is in the special attention that must be given to their deployment and operation. The fact that a mobile environment must make use of temporary and varying sites, services and other locally available resources such as labour and equipment means that great care must be taken in creating a workable scenario for deployment as well as use. This means that the operation of the building will always be different to its static counterpart with the same function, but it also often means that the delivery of the building from design inception to completion and operation has to be reconsidered. This has implications for every aspect of building provision from funding, contracts and staff training to the relationship with the public. Powerhouse::UK was a mobile exhibition building created by architects Branson Coates for the British government's Department of Trade and Industry that was designed to communicate the range and quality of British-made products. The building had to be made in a very short time-scale so it did not incorporate experimental construction techniques. However, it was very innovative in the provision of the building and its operation. In this case the designers became the creators and providers of a facility rather than a building. In collaboration with the client they decided the functions required and the way they could be provided, the appropriate imagery. They then engaged in the task of providing and managing every aspect of the project, which consequently became part building, part exhibition, part event. Every physical component became the designer's responsibility, from the temporary foundations to the guides' clothing. In addition, the same team controlled both the manufacturing and the operation of the facility. This led to successful delivery in a very short time-scale but also a consistency of image and message in the different elements of the event (Kronenburg, 2003, pp. 149–160).

Human response to technology is perhaps the most complex area that must be assessed in order to judge the success of architectural design – yet it is of course also the most important. People's response to architecture is subjective and individual. Though it is perceived as functional and useful it must also express the identity and aspiration of the communities for which it is created. There is no 'correct' image and character for architecture because it must always change in response to a wide range of conditions imposed by site, climate, culture and society. But architecture must also respond to its own making – be expressive of its source, both historical and technological. The character of a building made from stone is determined by its materiality – it has a language that can be understood by all. Today's industrial and technical prowess means that any material can be utilised in a way that bypasses its natural language. Photography, once seen as a representation of an authentic image, is now not trusted due to the commonplace use of digital photography which can be easily manipulated. In the same way, the language of building might become confused and unreadable if technology is not used honestly. London's Museum of the Moving Image hospitality pavilion was designed by architects Future Systems and engineer Peter Rice. It exemplifies technological innovation in the service of function rather than as applied image. Situated on a public pedestrian thoroughfare alongside the bank of the River Thames its appearance was both temporary and celebratory to communicate its role as a special event support facility. The technology used in its construction also had dual elements, familiar and innovative, that collaborate in a surprising manner. Clearly it is a tent, using a plastic membrane and the thin round rods familiar to all campers – but the form of the building and the way in which the materials were used give it the feel of a tent of the future: ultra lightweight in nature; both transparent and translucent; lofty and unconfining. This building shows that architecture can be temporary and refined, sturdy and elegant, functional and mysterious (Kronenburg, 2003, pp. 61–70).

III

Society is going through great change – social, economic, political and technological. The force of change is inevitable and the only constant is change itself. Like all aspects of modern life, architecture must respond to these new pressures if it is to remain useful and relevant. Building processes produce 50 per cent of all greenhouse gases. Changing the way we build by recycling materials, balancing embodied energy when selecting them, increasing overall building efficiency and, particularly appropriate for portable buildings, designing them for efficient and flexible use is imperative. Architecture must also respond to new regulations that deal with a wide range of new issues such as accessibility, safety and security. These demands must be met within the constraints of greater economic accountability. Buildings today are built to a higher specification, yet must also be built faster and with fewer personnel. They must now respond to lifecycle-costing assessments that deal with the impact they have on their users and the environment, and their efficiency over their entire life. Despite these pressures, the timeless qualities that architecture provides for society in terms of establishing continuity and purpose must remain. Transportable architecture is at the vanguard of innovation and therefore a resource for new ideas to which the rest of the industry can turn. It forms the testing ground for new materials, building techniques, delivery methods and operational strategies. But it also has special relevance because the nature of the problems it solves means that it must manifest itself in many diverse forms. In transportable architecture efficiency and operation are paramount and consequently success is measured by the careful balance of structure, space and operation. Image, meaning and beauty are important but these attributes must result more than usual from the functional demands placed on the design. It is therefore a form of architecture that is created outside the restrictions of aesthetic movements – it is a *zeitgeist* rather than a style.

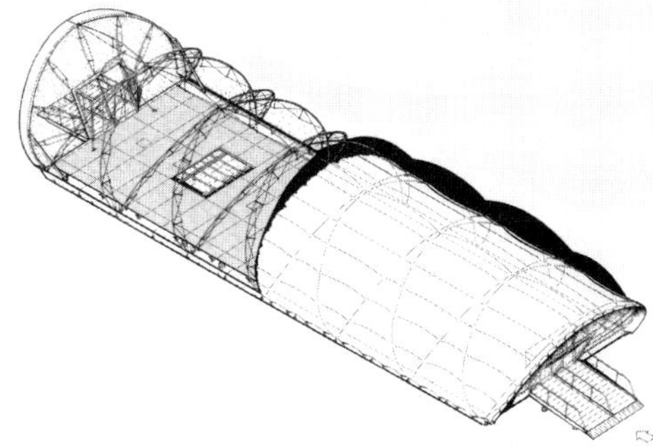

14.7 Future System's MoMI pavilion designed in conjuction with engineer Peter Rice (1994)

References

Horden, Richard (1995) *Light Tech: Towards a Light Architecture*, Berkhäuser Verlag, Basel.

Krell, David (ed.) (1993) Martin Heidegger, 'The Question Concerning Technology' in *Martin Heidegger, Basic Writings*, Routledge, London.

Kronenburg, Robert (2003) *Portable Architecture* (third edition), Architectural Press, Oxford.

Mellor, James (ed.) (1970) Buckminster Fuller, 'Designing a New Industry' in *The Buckminster Fuller Reader*, Jonathan Cape, London.

Material Innovations: Transparent, Lightweight, and Malleable

Filiz Klassen
Ryerson University

> We have only begun to speculate upon the uses of these new materials in architecture. Characteristic properties have begun to emerge however, in recently developed materials that are the opposite of many conventional materials now in widespread use. . . . Dramatic changes in the properties of recently developed materials will ultimately transform architecture. . . . Beyond infatuation [with new materials] . . . lies a world of purposeful form yet to be explored, a world in which materials will be selected based upon properties relevant to use. (Kieran and Timberlake, 2004, p. 121)

This chapter comments on the importance of new materials research through a brief examination of transportable or lightweight building components. In instances where the fact of portability and lightness becomes a form generator in architecture and design a different approach to available technology and material resources is required. The examples demonstrate that research on new materials and construction methods will continue to play an important role in the development of innovative design applications for transportable environments.

Transformation of materiality in architecture

Material innovations are as much a part of the current debate about architecture as they have been in any period in history. In western architecture and design over the last 150 years, the discovery of new materials such as titanium, synthetic polymers, and artificial ceramics or new applications for existing materials such as steel, concrete, glass, and paper have served to transform ideas of materiality from monolithic to ever more ethereal and ephemeral constructions (Beukers and Hinte, 2001, p. 13; Manzini, 1989, p. 107). In addition to formal and construction innovations, we have recently become more aware of how materials in architecture and design are extracted, cut, processed, and treated with high pressure or temperatures and how transportation to construction sites, utilization in a building, and the use or discarding of waste or debris at the end of the building's life cycle uses vast amounts of energy and simultaneously releases toxic substances. Sustaining natural and built environments by avoiding depletion of natural resources, lowering energy consumption during material production, employing recycling processes, and achieving higher performance of materials in their built state are forcing artists, designers, engineers, builders, and scientists to seek material innovations that go beyond what is conventionally available in the current building industry.

It is my contention that architects and designers in the present context of technological developments would benefit from the incorporation of what I provisionally call "material responsiveness" into their future practice of the design disciplines. With this term I suggest that architects and designers will develop a new relationship to materiality in general, moving away from the model of simple assemblage of traditional, pre-existing, and standard material choices, and the inherent conservatism of the construction industry at large, to be directly involved in the conception and development of new materials and material properties along with their applications into construction. This chapter examines selected material innovations through a review of current research and specific design case studies to foresee new directions in the available materials for the design of transportable environments.

New materials cannot transform the design and construction industry in a day, or even a decade, as it was imagined that the bold use of plastics would do in the 1960s and 1970s. New materials and methods will continue to coexist with what is currently available. Nevertheless, I believe that the range of materials available are in a state of transition in which current practices will inevitably be transformed in a direction compatible with more lightweight mobile construction.

Transparency

With technical advances in the production of large plates of glass from the beginning of the nineteenth century to today, glass – composed of silicates and an alkali fused at high temperatures – has been one of the most widely used construction materials. Current research into coating techniques, solar control technology, integration of microelectronic circuitry, and energy generation capacity with building integrated photovoltaics (BIPV) is giving glass a new place in construction innovation by aiming to eliminate the negative effects of heat gain or loss, glare and reflection, and sound transmission through the glazed surfaces (Weinstock, 2002a, p. 120). These innovations are radically transforming the design ideas of window and wall, transparency and opacity, and day lighting and shading in buildings.

Different kinds of glass have recently entered use in the construction industry: for example, photochromic glass that responds to light, and thermochromic glass that responds to heat. Although not yet widely used in architecture, photochromic glass is used in aircrafts, vehicles, appliances, and popular sunglasses and ski-goggles. Electrochromic technology changes the transparency of glass partitions or cladding from perfectly clear to translucent or completely opaque. This is achieved by passing low-voltage electrical charges across a microscopically thin coating on the glass surface that can be activated either manually with a switch or by sensors which react to light intensity. Thus glazing can control the level of solar transmission into the building, allowing minimization of heating or cooling and optimization of artificial lighting ("Electrochromic glass," 2003). Mostly used for privacy control in office settings, these windows have also been capable of sensational effects when used for the changing rooms in the Prada Store, New York, by the Office for Metropolitan Architecture (OMA).

In architecture, the transparency/translucency of glass or plastic is transformed into an active state by filtering light. It thus imitates the thermal resistance of an opaque material. Werner Sobek's R128 House proposes the use of light control technology to combine the transparency of windows with maximized energy efficiency for a complete building. The assembly, dismantling, and recycling of the house was built into the design process by choosing a lightweight, modular steel structure and glass cladding (Herwig, 2003, pp. 77, 79). In the effort to create a building shell that adapts itself to various levels of light transmission, absorption, and ventilation needs, Sobek proposes to integrate numerous mono-functional cells into glass that alter their chemistry to minimize energy input into the building (Blaser, 1999, pp. 59–63). His subsequent computer study for the proposed R129 House employs transparent cladding with a so-called intelligent plastic membrane coated with scratch- and chemical-resistant glass and solar cells to provide the majority of a building's energy supply. This thin electrochromic membrane is also capable of changing partially or entirely from opaque to transparent as there is no need for any individual material to be exclusively transparent in this energy-conscious environment.

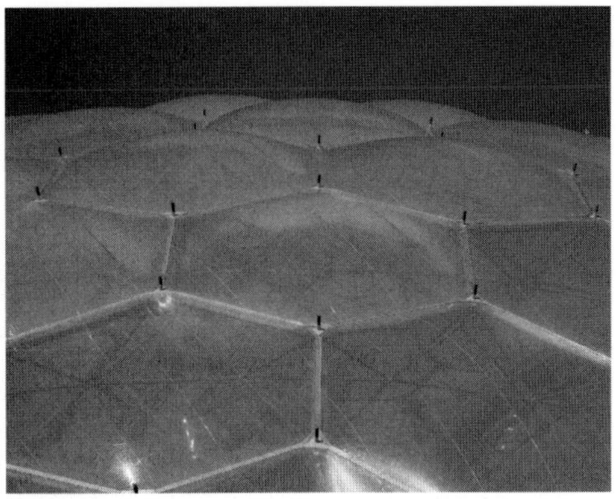

15.1 Eden Project, Nicholas Grimshaw & Partners

In buildings that require large areas of glass cladding or large span roofs, weight, limited span capabilities, and the expense of glass are deterrent factors. The qualities of new plastics such as transparency and energy efficiency make them attractive replacements for glass. For example, use of the advanced plastic ETFE (ethyltetrafluorethylene) pneumatic cushions as cladding for the Eden biosphere project (Figure 15.1) in the UK by Nicholas Grimshaw and Partners in 2001 demonstrates a number of desirable qualities. The inflatable three-layer pillows weigh only 1% of an equivalent glass cladding and the material efficiency of this system also allows a lighter supporting structural grid. Varying in size from 5 to 11 meters in diameter, hexagonal cushions were inflated immediately after being installed onto extruded aluminum perimeter frames with a permanent flexible tube feeding air into each cushion. The ETFE, extruded into a thin but resilient film, does not deteriorate as a result of polluted atmospheric conditions or ultraviolet light, and does not collect grime due to its low adhesive properties. Finally, it is environmentally friendly since few additives are used in its manufacturing process (Weinstock, 2002, p. 121).

The concept of SmartWrap™ (Figure 15.2) aims to revolutionize the building industry by using a single sheet of film as a building skin (McCormack, 2003). This concept, demonstrated at the Cooper Hewitt National Design Museum, experiments with the most common plastic material (polyethylene terephthalate, used in water bottles) to provide an overall cladding that eliminates the need to define permanently fixed transparent and opaque surfaces. SmartWrap™ incorporates emerging technologies in heating, cooling, visual display, lighting, and energy collection onto its surface through a printing process that is similar to that of inkjet printers. Climate control is achieved through micro-capsules of phase change materials embedded in a plastic cladding that provides latent heat storage for thermal moderation (i.e. absorbing, storing, or releasing heat as they change state). This installation also uses organic light emitting diodes (OLED), such as those used for mobile phone or palm pilot displays, deposited onto a glass or plastic substrate. Thin film silicon solar cells in the SmartWrap™ are used to power the OLED technology. This innovative cladding material is an ideal solar collector, moderates temperatures through phase change materials, and is able to provide lighting and information displays. This technology may become commercially available within a couple of years as a customizable printed façade that anyone can design at a computer terminal at Home Depot. It would arrive at the construction site within a week, ready to be applied (Kieran and Timberlake, 2003; McCormack, 2003; Northrop, 2003).

Lightweights

An unmistakable trend in architectural materiality is a movement toward the use of more lightweight construction materials. While the esthetic quality of lightness has become an attribute achieved through transparency of glass, the airy look of free-floating forms, planes, and minimalist detailing, these strategies do not necessarily generate a lighter-weight building or building component. The physical qualities of lightweight structures, however, tend to concentrate on structural ingenuity and better technical performance of structural members by the use of lighter material substances or structural compositions. For example, a space frame, honeycomb structure or fabric membrane is considered lightweight compared to a concrete beam or thin shell concrete structure since the comparable ratio of height, depth, and span is much less in the former than the latter. The goal of lightweight construction thus becomes the integration of material research and design application.

According to Ezio Manzini (1989, p. 89), "creating the light and resistant should be viewed as an area that does away with the traditional divisions of expertise between the chemist (who works with the

15.2 Smart Wrap, KieranTimberlake Associates LLP, exterior (left), detail (right)

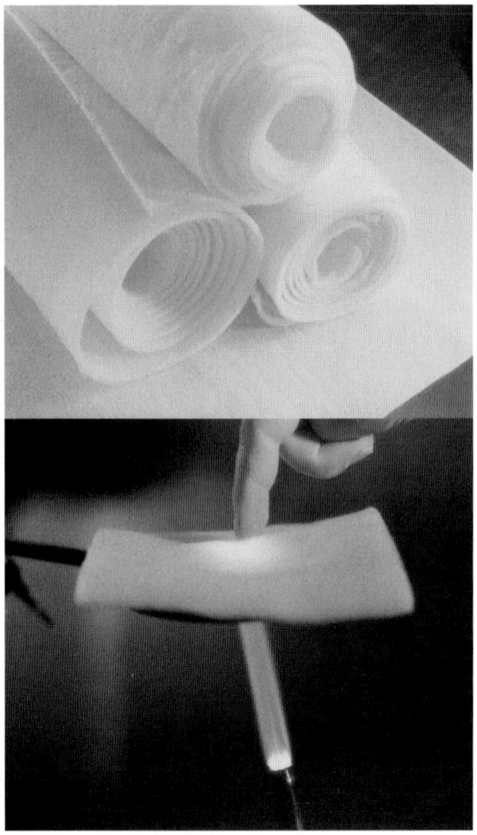

15.3 Aspen aerogel blanket

properties of materials) and the designer (who works with the form of the finished product)." He proposes that the object's form, the material macrostructure, and the microstructure of the individual material component must be resolved as a whole. Many new computer-aided design and manufacturing (CAD-CAM) processes allow and contribute to material developments in this area. Some materials offer extraordinary weight reduction in buildings in comparison with conventional materials of the same category. However, it must nevertheless be noted that lightweight is a relative term involving a comparison between suitable materials.

Developed by Steven S. Kistler in 1931, the world's lightest material, silica aerogel, known also as "frozen smoke" or "solid air," weighs only 3 milligrams per cubic centimeter (Wilson, 2003, pp. 6–7). The National Aeronautics and Space Administration (NASA) has used aerogel for insulation on its spaceships, on its Mars exploration vehicles and on the Stardust mission to collect particles from Mars' surface and debris of the "Wild 2" comet ("Aerogel," 2004). Aspen Aerogel, founded in 2001, produces aerogel in the form of a thin and flexible blanket that is used mostly in the aerospace industry and for insulation of space or ski suits (Figure 15.3) (Hogan, 2002). Cabot and Kalwall Corporations, leaders in specialty chemicals and construction, have together been looking into commercial applications for aerogel. They have recently developed insulated day-lighting panels with Nanogel™, a translucent aerogel. So far the commercial applications have been primarily the provision of transparent insulation. Panels by Kalwall Corporation are made with 19 mm-thick (0.75 inch) Nanogel™ and have the same R-20 insulation value as required for a standard house or building wall whilst still having about 20% translucency (Howe, 2002). The first architectural use of Kalwall's composite sandwich panel with Nanogel™ was over the swimming pool of a Comfort Inn Hotel in New Hampshire in 2003 (Solomon, 2003).

Honeycomb structures in architecture are known for their lightness and high strength-to-weight ratio. A number of companies produced composite materials that feature aluminum, cardboard, or synthetic honeycomb core sandwiched between two surface materials. Architects Mitman, Bourlier and Froech have developed a line of lightweight panels, named Panelite™ (Figure 15.4), that employs a honeycomb process used in the aerospace industry. This bonded honeycomb sandwich panel with translucent facings is a versatile building material providing a lightweight, self-structural panel for interior applications. Panelite™ offers varying degrees of visual privacy by transmitting, pixelating, coloring, or diffusing light. A comparative study based on a 1,200 by 2,400 by 19 mm (48 by 96 by 0.75 inch) panel shows that Panelite™, at 14.5 kg (32lbs), is 4.5 times lighter than acrylic and 10 times lighter than glass ("Panelite panels," n.d.). They arrived at various honeycomb cell geometries, such as hexagonal, over-expanded, quadrilateral, and tubular through an iterative-evaluative process that allows for experiments and errors with structural and light-transmitting properties of the materials used.

The Panelite™ Insulated Glass unit can be used on exterior façades and combines the transparency, durability, and weather-resistant nature of glass facings with directionally translucent, customizable properties of the tubular polycarbonate honeycomb core (Bourlier *et al.*, 2002, pp. 35–36). This provides an innovative solution for glazing and curtain wall applications such as that used at the Illinois Institute of Technology (IIT) McCormick Tribune Campus Center in 2003 by the Office for Metropolitan Architecture (Figure 15.5). Panelite's material development division is also devoted to developing other new materials in-house, ranging from mica laminates to structural fabrics (Ermann, 2003).

In relation to lightweight properties, composite materials (combinations of two or more materials with complementary properties) form another area

15.4 Panelite™ products. AO/FB (left): Aluminum honeycomb with blue fiberglass facing; PE/RX-O (right): Polymer honeycomb, colored resin facing

15.5 IIT McCormick Tribune Campus Center, OMA

15.6 Sketch of Eiffel Tower from knotted aramid fiber

of significant research in the search for stiff yet slim material properties. Currently, "composites are generally looked upon as compounds of polymers and other substances" (Beukers and Hinte, 2001, p. 70). However, they can be any combination of two kinds of materials, not only plastics that are usually different in nature. The result is a material that features the best characteristics and material properties of both. Probably the most common composite of all is reinforced concrete, mixing concrete and steel. Plastics laminated with wood form the oldest example of advanced composites. Most laminated countertops, for example, consist of layers of wood with one layer of phenolic plastic on top. Another common composite is fiberglass, a name that refers only to the fiber part of the complete composite material.

Carbon fiber reinforced plastics (CFRP) have stiffness and strength that can be twice that of steel, with only one-quarter of its density. Development of the use of CFRP in the aerospace sector is one area where demand for structures with high strength and elasticity, particularly when used in high-temperature environments, is evident. Since its commercialization in the mid-1970s, the carbon fiber industry is promoting accessibility of CFRPs beyond aerospace applications to improve the quality of mass-produced items such as cars, construction components, and sports equipment. By changing the fibers, CFRP has been developed to reduce its cost over time (Bucquoye, 2002, p. 24). Recently, Testa and Weiser (TESTA Architecture and Design) have undertaken the design of a high-rise tower using composite materials, particularly carbon fiber with complex computer modeling tools.

Perhaps the lightness that composites offer in architectural construction has not been given the credit it deserves due to the material's unfamiliarity and ecological issues that arise in its production processes and recycling (Bucquoye, 2002, p. 165; Solomon, 2003). The formulations of resins and fibers, though strong, are expensive to produce and use petroleum and other non-renewable resources in their production. Nevertheless, composites continue to be promoted through exhibitions and innovative design work to raise awareness of these materials through a hands-on method and illustration of their diverse applications which range from consumer goods and aerospace to art ("Definition of composites," n.d.). Marcel Wanders set carbon and aramid fibers, known for their stiffness, in resin to make his well-known "knotted chair." His interest in these fibers extends beyond furniture to projects for large-scale structures, such as his plan to build a replica of the Eiffel Tower (Figure 15.6) directly adjacent to the original tower. He calls this conceptual project a virtual collaboration with Gustave Eiffel. From a helicopter, Wanders intends to hang an "A frame" made from knotted aramid rope and impregnated with epoxy with the same general shape as the Eiffel Tower. Flexible at first, the frame will harden within a matter of hours and gravity will give Eiffel 2 its final form.

A recent trend in lightweight materials combines biotechnology and composite materials. Composite materials from renewable plant sources made their first appearance in January 2003 in the Model U Car by Ford. It has soy-based body panels, seat foam and grease, as well as corn-based tires and sunflower-seed-based engine oil (Wilson, 2003). The University of Delaware's Affordable Composites from Renewable Energy Sources (ACRES) Group is experimenting with various resins and fibers based on plant sources such as soy, corn, and hemp; they expect these bio-based structural composites to appear shortly in building construction. Lightweight bio-based foam from soy oil might replace commercially available foam. Flax and hemp fibers could offer a sustainable and renewable source of raw materials for a range of industries. "Potential uses of the fibers developed from 'non-food crops' range from replacement for fiberglass in composites for the automotive industry to short-fiber flax suitable for cotton spinning, to produce high value textiles" ("Sustainable composites," 2003).

Nexia Biotechnologies Inc., Montreal, Canada, develops and manufactures complex recombinant proteins for use as biomaterials and biopharmaceutical products with mostly medical and industrial applications (Nexia Biotechnologies, n.d.). Their Biosteel™ product is obtained from genetically engineered goat's milk that has spider-silk protein and is the strongest fiber yet produced. Spider silk is five times as strong as steel with much higher tenacity. Turner, head of Nexia, thinks that in the far future, rather than lifting objects into orbit by space rockets it will be possible to make a space elevator stretching to extraterrestrial locations using Biosteel (Newman, 2003, p. 70).

Although composite materials are replacing metals, skeletal steel frames, and metal cladding as the lightest structures in architecture available since the beginning of the twentieth century, Annette LeCuyer comments that extremely light alloys, technology, and digital processes transferred from other industries are changing the use of metals in building construction both in the design applications of structural frames and building skins (LeCuyer, 2003, pp. 7–15). Despite their high cost, high-strength and light metal alloys of aluminum, magnesium, and titanium offer advantages such as being able to be formed at lower temperatures and providing higher strength and toughness in comparison to typical carbon steels. Although the presence of some of these metals on the earth is abundant, extracting them proves to be an energy-intensive and costly process. These lighter-weight metal alloys have found application in compact laptops, household electronics, aerospace applications, surgical implants, and prosthetic devices, but construction use to date remains limited and costly.

Foamed metals, at densities less than one-tenth that of water, also present ultra-lightweight, cost-effective, rigid, and fire- and impact-resistant construction alternatives. Previously confined mostly to the automotive manufacturing industry, foamed metals offer uses as sandwich beams and building panels that primarily use aluminum foam as a core in contrast to structural foams based on polymers ("Metal foam," 1998). Alusion™ (Figure 15.7) is a 100% recyclable panel of stabilized aluminum foam (SAF) that combines the lightness of foam with the strength of the metal (a 25 mm-thick (1 inch) bench is capable of bearing 1,814 kg (4,000 lbs)). It is currently used for wall and ceiling panels and for furniture but has the potential for far wider use in the future (Onna, 2003, p. 73).

15.7 Alusion™

Malleable matter

Recent innovations in fabrics and in fiber composites that have already impacted on industrial design, fashion, furniture, medicine, and aerospace engineering are likely also to advance a transformation in people's perceptions of personal space and the physical boundaries of the built environment (Braddock and O'Mahony, 1999, pp. 6–7). The term "flexible in design" embodies the concept of malleability of matter, form, and space. It emphasizes the stretchable, expandable, or contracting qualities of construction. Fabrics and textiles are the most common form of malleable matter and have been used as components in the built environment in the form of nomadic tents, awnings, or temporary structures used for shade since mankind first began to build shelters.

Fabrics, considered to be the fifth category of building materials (following stone, wood, metal, and glass), are most often made from either natural fibers such as wool, cotton, and silk or their synthetic counterparts. Some synthetic fibers are nylon, polyester, high-strength carbon fibers called aramids, and high-temperature-resistant fibers such as Nomex and Zylon (Newman, 2003, p. 57). Some engineered textiles in construction are PVC

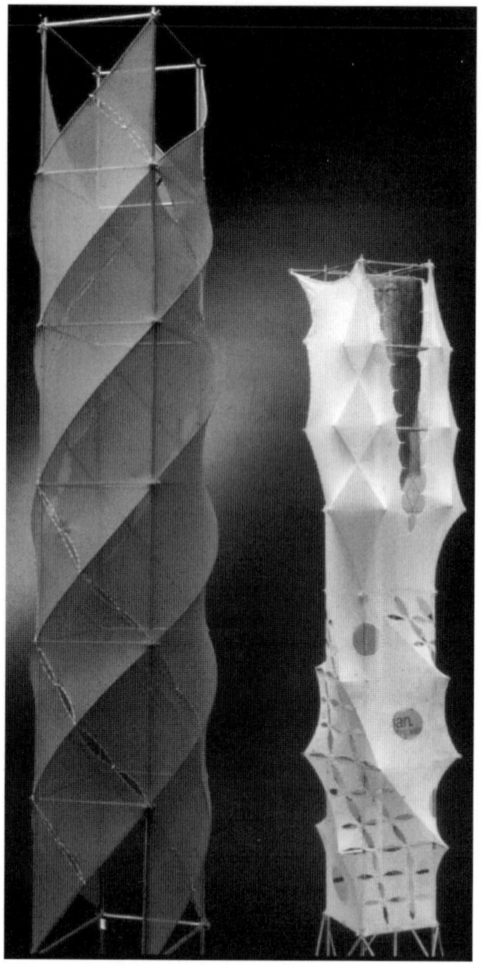

15.8 Fabric skyscraper

(polyvinyl chloride) and ETFE (ethyltetrafluoroethylene) foils. New developments in engineered textiles that are used in tensile or inflated structures (with coatings such as coating PTFE (polytetrafluoroethylene)) and embedded technology have brought fabrics to the forefront of lightweight and portable architectural applications as replacements for heavier construction materials. The strength of inflatable high-strength fabrics is being tested in fabric beams, columns, and heavy cargo lifting that goes beyond the architectural applications of large-span structures and inflatable portable buildings. Pressurized air beams provide an ability to rapidly erect large and lightweight structures such as aircraft hangers. Air beams are manufactured by braiding a high-strength three-dimensional Vectran fabric over an air-retention "bladder" (Taylor, 2002–03). The Cargolifter Airship hangar, designed by SIAT Architektur and Design and Arup Engineering, is noted as the world's largest free-standing aircraft hangar spanning 225 meters with steel arches covered by a woven fiberglass substrate with Teflon PTFE-coated roofing membrane.

FTL Design Engineering Studio, New York, has been an innovator in the use of textiles in architecture for the past twenty-five years. Since the early 1990s, they have been exploring the feasibility of cladding a skyscraper with tension fabric membranes, an innovative reinterpretation of conventional glass curtain walls (Figure 15.8). The project anticipates the use of double- and triple-skin fabric walls approximately 5 by 30 meters that are stressed with heated airflow to provide both structural rigidity and a thermal barrier at a much lower weight and cost than conventional materials (Watt, 1995, p. 49).

The integration of graphics, luminosity, and energy-efficient photovoltaics into fabrics presents designers with a challenging innovation area. FTL worked on the Cooper Hewitt National Design Museum's exhibition in 1998 entitled *Under the Sun*, designing the first tensile structure with integrated photovoltaics (Willmert, 2002). This solar pavilion used a woven fabric with integrated photovoltaic technology as a structural skin that transformed sunlight into power. Thin-film amorphous silicon photovoltaic modules just 0.06 mm (0.005 inches) thick were encapsulated and laminated into the woven fabric. The versatility of this system can be seen in a recent lightweight, energy-conscious design project entitled *Powershade* that integrates Iowa Thin Films PowerFilm™ flexible solar panels directly with the tent fabric (Figure 15.9). This mesh fabric can either be used to cover an existing tent or as a standalone structure, and can be used as a shelter/power generator tent or for temporary applications in recreation, military, and disaster relief efforts.

FTL's *Advanced Inflatable Airlock* (AIA) is a research and development project in support of NASA's Space Launch Initiative (Figure 15.10). The airlock functions as a temporary pressurized structure to allow astronauts to leave the space vehicle and perform extra-vehicular activities (EVA) or "space walks." The research involves making the airlock out of many layers of fabric, instead of heavy aluminum which is the current technology. Currently this research project is in the technology development stage and if selected for further development, the AIA may be in space in five to ten years from now as part of NASA's next-generation space shuttle.

Also concentrating on experiments that create a flexible architectural skin, Kas Oosterhuis of ONL Netherlands considers architecture as a "lean and flexible construct." Their programmable space called the *Muscle* (Figure 15.11), for their interactive pavilion at the Non-Standard Exhibition at the Centre Pompidou in Paris, proposed a pressurized soft volume capable of changing its shape by contracting and relaxing pneumatic muscles designed and manufactured by Festo KG. The individual muscles were orchestrated with the use of sensors and

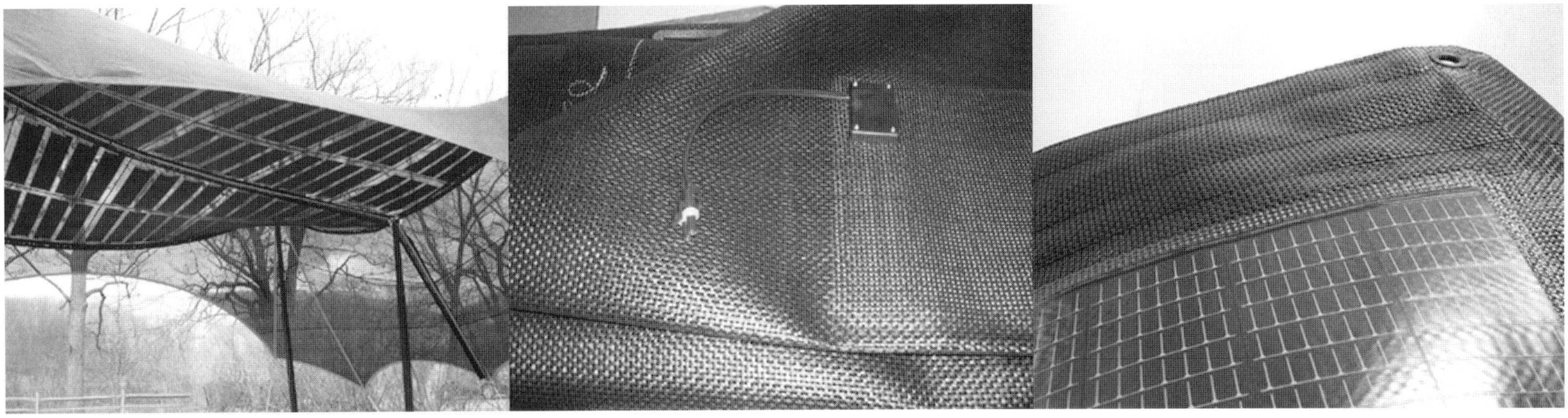

15.9 *Powershack*

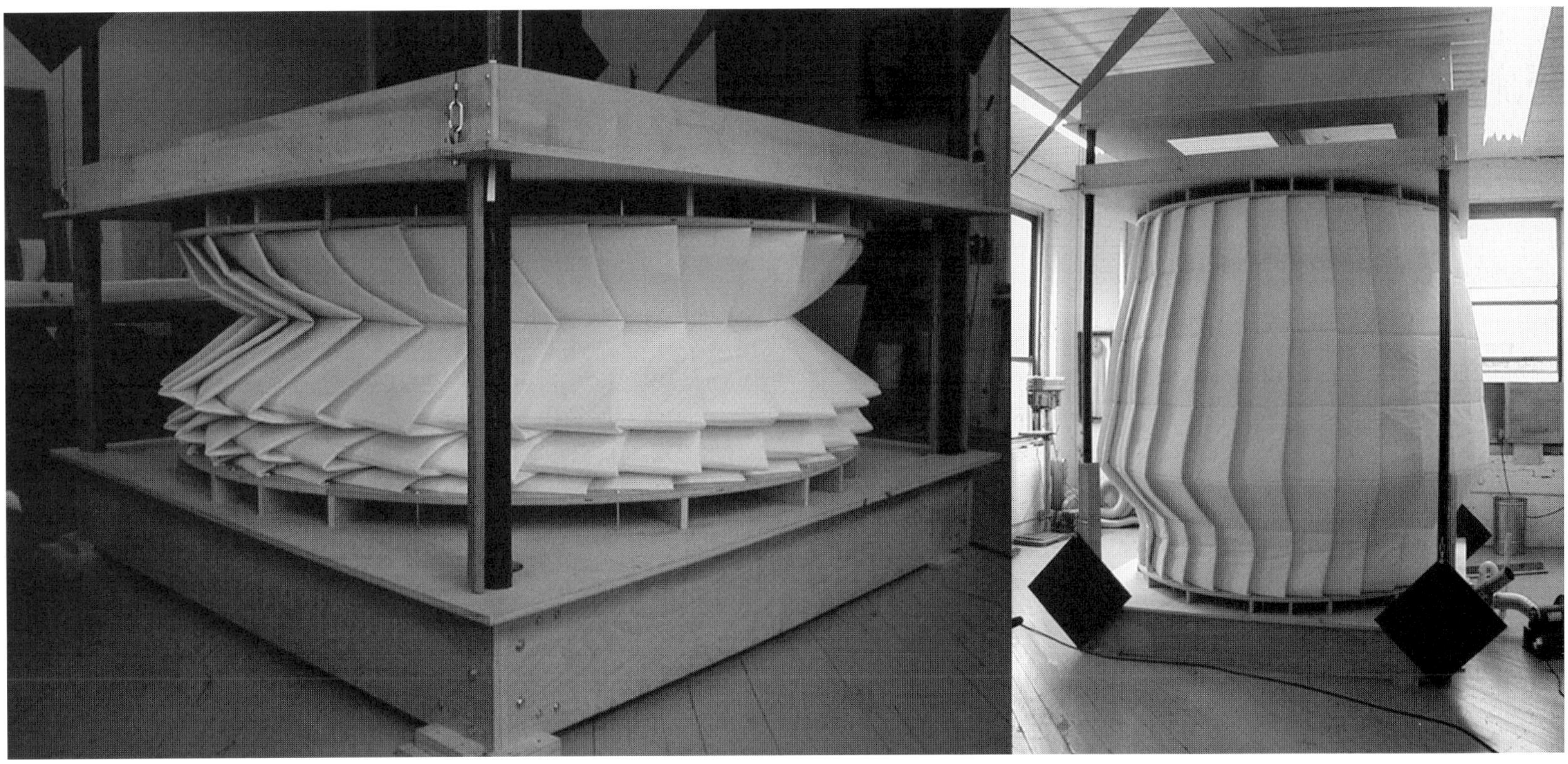

15.10 *Advanced Inflatable Airlock* (AIA)

15.11 *Muscle*

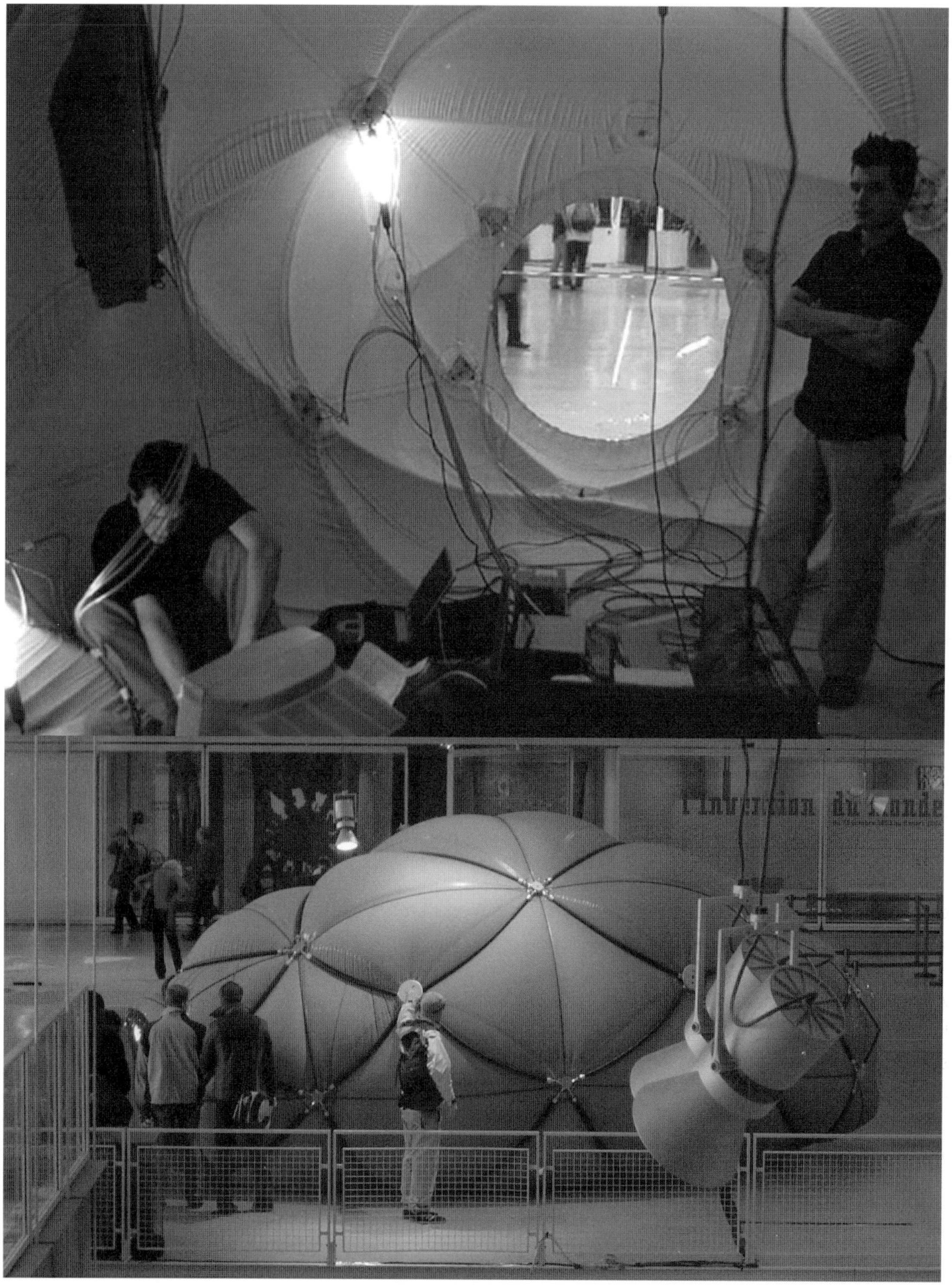

embedded computer technology to change the length, height, width, and thus the overall shape of the space when triggered by people passing by or touching computer screens. In architecture, a change in construction materials' properties (their elasticity or volume) in response to a change in the environment is usually considered to be a potential problem – something to be avoided. However, these experiments demonstrate that innovations in the area of flexible architectural skin depend on taking advantage of pliability or malleability of fabric material to emphasize the changing form.

Conclusion

We can no longer afford to design objects, buildings, and spaces without regard to the inevitable changes in function, reuse, and/or rebuilding over time. In architecture and design, size, weight, and longevity matter. As designers we should downsize, make things more lightweight, and offer increased transformability to prolong the life span of built spaces, objects, and materials. A change in attitude of both designers and consumers should emphasize transportable and transformable design, not as a one-off design solution, but as an integral part of the design and production processes. At the same time, the creation of lightweight and transformable construction elements in architecture and design can remain compatible with stability and longevity – qualities that are more commonly associated with weight and fixity.

A transformation of design attitudes emphasizes the importance of research on new materials and construction methods as well as new uses for existing materials. Considering material science and technology transfers from other industries as instigators for design, Kieran and Timberlake see new materials as opportunities to save weight and add strength and durability to construction whilst lowering costs and economizing on construction time. They hope that:

> there will be regular affiliations and alliances with material scientists and product engineers, working together as models of collective intelligence, making large parts of buildings in high quality, controlled settings, using materials they're not using now, purposeful materials, not just collections of neat-looking materials.
> (Ermann, 2003)

Kennedy and Violich (KVA) is another progressive firm which is exploring the integral relationship between light, information, thermal control, and material properties through their working group on the design and research of new materials. Sheila Kennedy states that "the architectural imagination is well suited to take on interdisciplinary problems and coordinate strategies for idea production and fabrication in architecture" (Mori, 2002, p. 11). Peter Testa, the founding director of the Emergent Design Group (EDG) in the Department of Architecture and Artificial Intelligence Laboratory at the Massachusetts Institute of Technology, formulates the main goal at EDG as research into structural morphology and new spatial models with a focus on emergent properties of material forms in architecture. A secondary goal is to develop simulations, tools, prototype designs, and building systems that test these principles in practical applications. The projects at EDG initiate new developments by combining innovations in modeling theory, intelligent systems, organizational theory, and the science of dynamics to transform the way designers use a range of currently available technologies (Testa, n.d.; Knecht, 2003).

Scientific research has produced materials that last longer, reduce waste, change form, and adjust to environmental conditions in different contexts. Architects and designers need knowledge, ethics, and creativity to transfer these technological innovations, often developed by other industries, into the built environment. Achieving or expressing absolute truth of design through autonomous artistic organization and material embellishments no longer seem to provide a convincing goal for many innovative architects. Many architects are becoming involved in the "alchemy" of construction, translating advanced material innovations into a more responsive and responsible built environment.

> Contemporary designers are facing the challenge of defining . . . [materials'] new multifaceted manifestations. The mutant character of materials, as expressive as it is functional and structural, generates new forms and a more experimental approach toward design.
> (Antonelli, 1995, p. 17)

References

"Aerogel" (2004) Retrieved on February 10, 2004, from http://stardust.jpl.nasa.gov/tech/aerogel.html.

Alusion, technical specification (n.d.) Retrieved on February 9, 2004, from http://www.alusion.com/specsheet.pdf.

Antonelli, P. (1995) *Mutant Materials in Contemporary Design*. New York: MOMA, distributed by Abrams.

Ashby, M. and Johnson, K. (2002) *Materials and Design: The Art and Science of Material Selection in Product Design*. Oxford: Butterworth-Heinemann.

Beukers, A. and Hinte, E. V. (2001) *Lightness: The Inevitable Renaissance of Minimum Energy Structures*. Rotterdam: 010 Publishers.

Blaser, W. (1999) *Werner Sobek: Art of Engineering*. Basel: Birkhauser.

Bourlier, E., Mitman, B. C., and Froech, A. (2002) "Material effect: a non-alphabetical glossary of the cellular-core translucent panel" in *Architectural Design*, September–October, 72(5), pp. 34–39.

Braddock, S. and O'Mahony, M. (1999) *Techno Textiles: Revolutionary Fabrics for Fashion and Design*. New York: Thames and Hudson.

Bucquoye, M. (ed.) (2002) *Van Bakeliet tot Composiet: Design met Nieuwe Materialen, (From Bakelite to Composite: Design in New Materials)*. Ghent: Design Museum, Stichting Kunstboek.

"Definition of composites" (n.d.) Retrieved on February 9, 2003, from http://www.mtm.kuleuven.ac.be/Research/C2/comptour/.

"Electrochromic glass" (2003) Retrieved on February 11, 2004, from http://www.glassonweb.com/glassmanual/topics/index/electrochromic.htm.

Ermann, L. (2003) "Technology transfer remains a nascent movement but more architects take up the challenge" in *Architectural Record*, 03(10), pp. 1–3. Retrieved on November 27, 2003, from http://archrecord.construction.com/innovation/3_BuildingTech/0310TechTransfer.asp.

Hart, S. (2002) "There is no north arrow in outer space" in *Architectural Record*, 02(12), pp. 1–8. Retrieved on December 5, 2003 from http://archrecord.construction.com/resources/conteduc/archives/0212space-1.asp.

Hart, S. (2003) "Imagining the future: How will we make buildings in 2030?" in *Architectural Record*, 03(10). Retrieved on December 5, 2003, from http://archrecord.construction.com/innovation/2_Features/0310future.asp.

Herwig, O. (2003) *Featherweights: Light, Mobile and Floating Architecture*. Munich: Prestel.

Hogan, J. E. (2002) "Technology." Retrieved on September 1, 2004, from http://www.aerogel.com/technology.htm.

Howe, J. P. (2002) "The light stuff: Cabot process allows commercial use of 'aerogels'" in *The Boston Globe*, December 27. Retrieved on December 10, 2003, from http://www.kalwall.com/nano5.htm.

Kalwall and Nanogel. (n.d.) Retrieved on December 10, 2003, from http://www.kalwall.com/nano1.htm, http://www.kalwall.com/news-insulation.htm, and http://w1.cabotcorp.com/controller.jsp?entry=product&N=23+1001+4294966858.

Kieran, S. and Timberlake, J. (2003) "Transfer technologies." Retrieved on December 6, 2003, from www.mb2010.com.research/transfer_tech.html.

Kieran, S. and Timberlake, J. (2004) *Refabricating Architecture: How Manufacturing Methodologies are Poised to Transform Building Construction*. New York: McGraw Hill.

Kieran Timberlake Associates LLP (2003) *SmartWrap™: The Building Envelope of the Future, a Mass Customizable Print Facade*. Philadelphia: Becotte and Company.

Klassen, F. (2003) "The Malleability of Matter" in *On Site Review*, 10, p. 48.

Knecht, B. (2003) "Brave new solid-state, carbon-fiber world: Architects Peter Testa and Sheila Kennedy are reinventing the design process through collaboration with industry" in *Architectural Record*, 03(10). Retrieved on December 5, 2003, from http://archrecord.construction.com/innovation/2_Features/0310carbonfiber.asp.

Kronenburg, R. (1997) *FTL: Softness, Movement and*

Light. London: Academy Editions. Architectural Monographs No. 48.

LeCuyer, A. (2003) *Steel and Beyond: New Strategies for Metals in Architecture*. Basel: Birkhäuser.

Manzini, E. (1989) *Material of Invention*. Cambridge, Mass.: MIT Press.

McCormack, J. (2003) "The cleverest building material around" in *Metropolis*. Retrieved on December 6, 2003, from http://www.metropolismag.com/html/sustainable/case/smartwrap.html.

"Metal foam delivers superlight components' (1998) (September) Retrieved on March 2, 2004, from http://www.eureka.findlay.co.uk/archive_features/Arch_Materials/METFOAM/metfoam.htm.

Mori, T. (ed.) (2002) *Immaterial/Ultramaterial: Architecture, Design, and Materials*. New York: Harvard Design School in Association with George Braziller Publisher.

Newman, C. (2003) "Dreamweavers" in *National Geographic*, January, 203(1), pp. 50–73.

Nexia Biotechnologies (n.d.) Retrieved on December 12, 2003, from http://nexiabiotech.com/en/00_home/index.php.

Northrop, J. (2003) "Inaugural exhibition solos: SmartWrap features building skin of the future." Cooper-Hewitt, National Design Museum press release, August 4. Retrieved on December 6, 2003, from http://ndm.si.edu/SOLOS/.

Onna, V. E. (2003) *Material World: Innovative Structures and Finishes for Interiors*. Amsterdam and Basel: Frame and Birkhauser Publishers.

"Panelite panels" (n.d.) Retrieved on November 7, 2003, from http://www.e-panelite.com/_downloads/PANELITE_BROCHURE.PDF.

Schulte, K. (ed.) (2000) *Temporary Buildings: The Trade-fair Stand as a Conceptual Challenge*. Corte Madera, Calif.: Ginko Press.

Slavid, R. (2002) "Inside Germany's space capsule: The world's largest free-standing aircraft hangar has been built near Berlin using five steel arches that span 225 m" in *Metal Works*, Spring, pp. 14–15.

Solar tensile pavilion (1998) "Cooper-Hewitt, National Design Museum, Under the Sun: An Outdoor Exhibit of Light." Retrieved on February 9, 2004, from http://ndm.si.edu/EXHIBITIONS/sun/start.htm.

Solomon, B. N. (2003) "Architects slowly begin to expand the traditional palette of materials: New substances from high-tech laboratories enter the realm of construction" in *Architectural Record*, 03(11), pp. 1–8. Retrieved on December 10, 2003, from http://archrecord.construction.com/resources/conteduc/archives/0311edit-1.asp.

"Sustainable composites" (2003) *Future Materials*, December. Retrieved on January 30, 2004, from http://www.inteletex.com/FrontPageFeatures.asp?PubId=28&NewsId=2471.

Taylor, B. (2002) "Inflatable structures: Air apparent" in *Tents*, December–2003 January. pp. 48–51. Retrieved on January 10, 2004, from http://www.ifai.com/File.php?ID=509.

Testa, P. (n.d.) "Research." Retrieved on February 19, 2004, from http://mit.edu/arch/edg/#.

Topham, S. (2002) *Blowup: Inflatable Art, Architecture and Design*. Munich: Prestel.

Watt, S. J. (1995,) "Fabric skyscrapers" in *Architectural Design*, September–October, No. 117, vol. 65 (9/10).

Weinstock, M. (2002a) "Engineering exegesis: Transparency and the Performance of Building Skins" in *Architectural Design*, September–October, 72(5), pp. 120–123.

Weinstock, M. (2002b) "Engineering exegesis: soft materials strong structures" in *Architectural Design*, March–April, 72(2), pp. 119–124.

Willmert, T. (2002) "Architects discover the flexibility of lightweight and durable fabrics: advances in films and coatings make fabric an excellent alternative to glass" in *Architectural Record*, April, 02(04), pp. 1–7. Retrieved on February 5, 2004, from http://archrecord.construction.com/resources/conteduc/archives/0204Fabrics-1.asp.

Wilson, A. (2002) "When most clothes will talk" in *Future Materials*, May–June, p. 9.

Wilson, A. (2003) "Aerotech: Solid Air, the world's lightest material has just become lighter still" in *Future Materials*, January–February, pp. 5–7.

Mobilized Manufacturing: The On-site Construction of Freeform Composite Shells

Jordan Brandt and Alejandro Ogata

Introduction

Tents, trailers, and their multiple permutations are the prime examples of "Transportable Environments" (TEs). They adhere to the concept of a shelter in a particular configuration that can be moved from one place to another with relative ease, usually when accommodations at the specified destination are unavailable or impractical. TEs have benefited greatly from technological developments in the fields of computer-aided design/computer-aided manufacturing (CAD/CAM) and materials research, increasing their efficiency (size, weight, durability) and adaptability. Using this same technology, it is now possible to provide the benefits of an entire manufacturing facility in a compact container, effectively integrating the core components of TEs: efficiency, adaptability, and mobility. This chapter speculates on how such a system might operate and describes some early experiments in creating a practical working model to prove the concept.

The mobile manufacturing unit (MMU) proposed here utilizes only three basic building materials: reinforcement fabric, core material, and resin, which are fabricated into monolithic thermal shells through the use of specialized computer software and custom CAM equipment. These raw materials have a high compressibility (i.e. a large constructed surface area relative to shipping volume) and readily conform to almost any shape. When combined, the resulting composite sandwich yields a high strength/weight ratio, rigid structural performance, thermal protection, and minimal material interfaces (joints). This composite sandwich shell can be fabricated as one large envelope or any number of discrete structural components on a variable form mold. With a formal vocabulary ranging from flat panels to synclastic curvature, it is possible to simply ship the MMU to the site and then address issues such as climate and terrain without attempting to predict and accommodate all potential variables (Figure 16.1). Furthermore, the MMU can be utilized to make any modifications to the structure after erection, including future additions and repairs.

Materials

The fundamental material categories of reinforcement fabrics, cores, and resins define three families of specific products with a diverse range of properties. Products within each family can be selected to optimize the composite sandwich for specific applications. Therefore the selection criteria must include not only performance characteristics (compressibility, weight, etc.), but also cost, environmental impact, and compatibility with the other sandwich components. With these parameters, the following analysis has defined a small group of materials that demonstrate promising characteristics for mobile manufacturing.

Reinforcement fabrics

These are combined with a matrix resin to provide the interior and exterior laminates of the structural shell. The laminates are structurally analogous to the flanges of a typical wide-flange steel beam, so the orientation and placement of the fabric can be optimized to direct the load paths and create a more

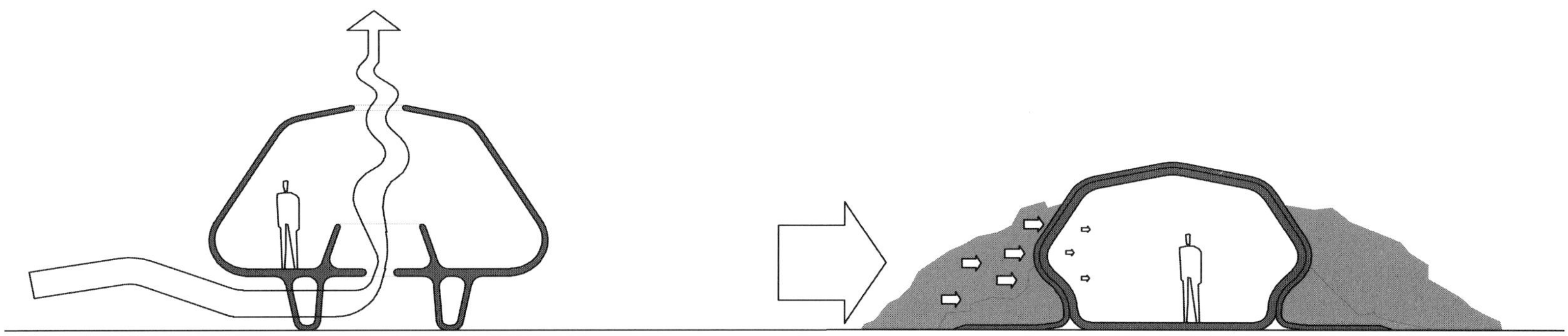

16.1 Multiple design permutations

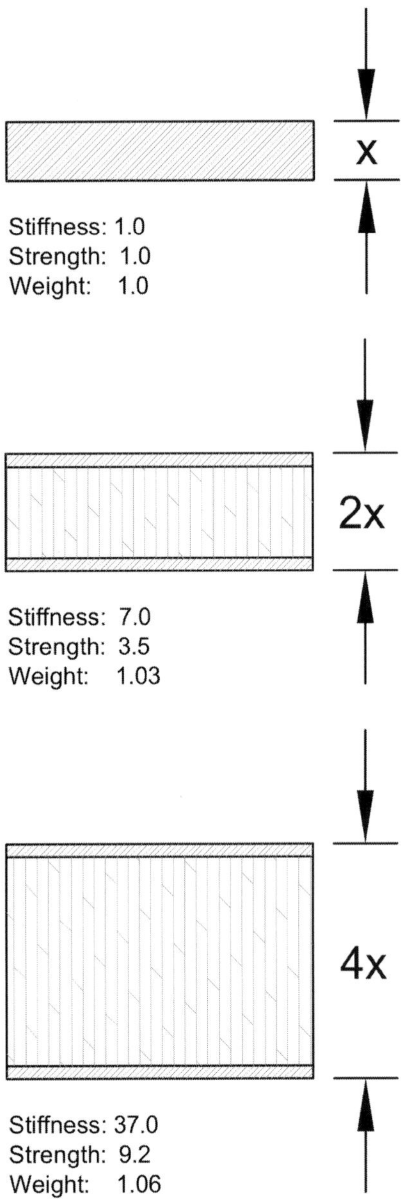

16.2 Sandwich structure strength/weight/thickness

efficient structural member. The structural capacity of the laminate is directly linked to its bond with the resin and to the core material. Through experiment it has been discovered that even small "bubbles," air pockets between the laminate and core material, can result in premature buckling and therefore failure of the entire panel.

Carbon

Carbon fiber has become virtually synonymous with the term "advanced composites," because of its high modulus (stiffness), high tensile strength, and corrosion resistance with a low relative weight. Despite these attractive features, carbon fiber has not seen widespread use in the architectural field due primarily to high material cost and an historical apprehension about funding the research to implement new materials. Industries with higher profit margins, however, have a vested interest in funding material research and development costs, because an increased performance yields higher profits (e.g. lighter airplanes = reduced operating costs) when viewed upon a scale of mass manufacturing. Likewise, one MMU could mass manufacture custom carbon architecture at a reduced total cost because the material performance benefits of transportability and lightness mitigate shipping costs and construction time.

Glass

Fiberglass is the cheapest and most common composite material in use today. From boat hulls to bathtubs, fiberglass has been optimized for a variety of scales and applications. Exterior architectural applications have exploited the pliability of fiberglass by mass manufacturing a diverse set of molded prefabricated components, while long-term exposure has proven the durability of fiberglass components. "E-glass" fibers comprise the majority of the fiberglass industry, providing a cheap and relatively strong building material that is available in either random-oriented strand mats or fabrics of various weaves. The more expensive S-glass fibers offer a multitude of performance benefits such as higher strength, modulus, and temperature values.

Natural

Over the past ten years, natural fibers have gained popularity as a replacement for glass. This new interest is fueled by a higher stiffness/weight ratio as well as potential lower costs and a lower environmental impact. There are, of course, some drawbacks to natural fibers that need to be overcome in order for them to be a truly viable option. Natural fiber has lower durability and impact strength when compared to its glass counterpart. However, tests (conducted on natural fibers in resin transfer molded composites) have demonstrated that fibers such as jute have only a minimal disadvantage when compared to glass in flexural strength, environmental exposure, and surface erosion (O'Dell, 1997).

Cores

The core material in a sandwich construction carries compression and shear forces to allow the laminates to carry the bending forces. The thickness of a core material exponentially dictates the stiffness of the structural sandwich (Figure 16.2), and Strong (1989) notes that sandwich construction provides "the highest stiffness to weight ratio of any common materials design." (p. 173).

For the purposes of the MMU, the core material must be compressible for transportability and load bearing when expanded. The core must also conform to the desired geometry in its expanded state without resulting in residual stresses that could potentially distort the panel when it is removed from the mold. The answer to the conundrum of flexibility, rigidity, and expandability lies in the proper synergy of physical and geometric properties.

A multitude of materials and designs are currently in use by the composites industry, but the presented parameters of the MMU drastically narrow the selection. Thermoform cores are extremely pliable when heated and provide adequate thermal and mechanical properties at low cost, but they are not expandable. Conversely, spray-on polyurethane foams show great potential in their expandability, but obtaining a uniform thickness presents an issue. The research described in this chapter focused upon paper honeycomb cores due to low cost, availability, and compatibility with the presented design parameters.

Honeycomb

With long-time use in many industries, cardboard honeycomb offers a lightweight, low-cost, and compressible core material. There are two primary modes of manufacturing: expansion and corrugation processes (Figure 16.3). The expansion process begins with sheets of paper and alternating lines of glue that are adhered to each other and expanded, resulting in hexagonal-shaped cells. The corrugation process first forms the sheets with a roller, and then sticks these sheets together providing a honeycomb pattern that is already expanded. It was found that the compressed core obtained from the first (expansion) process is difficult to expand, and results in residual stresses that can distort the panel after fabrication. This considered, the corrugated approach appears to be preferable, whereby the core is shipped in bundled sections that expand when opened, thus negating the need for specialized core expansion equipment.

Resins

"The matrix (resin) in a composite can be thought of as performing two major roles. First, it transfers loads to the reinforcement. Secondly, it protects the reinforcement from adverse environmental effects" (Strong, 1989, p. 9). The resin system binds the yarns of the reinforcement fabric to each other and the core, unifying disparate components into an effective structural sandwich as well as surrounding the individual fibers with a protecting shell. Resins subdivide into two general categories, that of thermosets and thermoplastics. Thermosets typically consist of liquid resin and a liquid initiator, which induces molecular cross-linking producing a homogenous polymer. Conversely, thermoplastics are non-cross-linked solids that are heated to achieve shape conformability and cooled to a rigid form. The temperatures required by the thermoplastics process are in the range of 370 degrees Celsius (700 degrees Fahrenheit), thus requiring large and heavy melting ovens inappropriate for the MMU.

Of the three materials required by the MMU, resins pose the biggest problem in terms of storage, transportation and use. Petroleum-derived resins (solvent-based) are usually high volatile organic compounds (VOC) emitters and hazardous to the environment. In fact, as experimentation progressed it was only possible to lay up resin-rich fabric for 1.5 hour intervals due to dizziness. In an attempt to overcome some of these negative factors, natural resins were researched for potential application. Recent developments in natural resins have focused on the promise of soybean oil. Currently the cost of soy-based resin is higher than solvent-based resin, but as technology improves (for example by genetically engineering a product of consistent quality regardless of season or growing area) prices will diminish accordingly. Furthermore, the possibility of combining natural resins with natural fibers could allow for the fabrication of a composite sandwich made only from renewable materials, fully recyclable, and environmentally benign.

Design and construction

By coupling a digitally integrated design, manufacturing, and erection process with the described

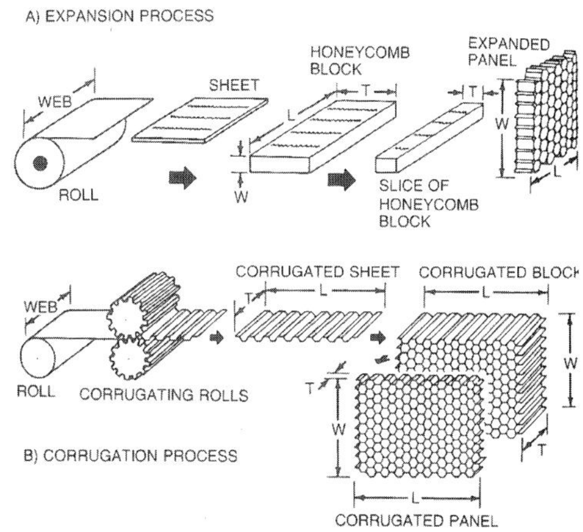

16.3 Honeycomb, expansion, and corrugation processes (Society of Manufacturing Engineers, 1989)

monolithic structural/thermal system, the MMU potentially provides a simple method to produce intelligent architecture. The process is initiated by obtaining global positioning satellites (GPS) coordinates of the site, which become the datum by which all subsequent geometry is created in the parametric model. If these site coordinates are updated, the entire geometry adjusts. In this way, multiple shelters can be designed for variable topology by simply providing new location data. The parametric model then directly feeds the CNC fabrication equipment and erection procedures for the composite panels. This seamless process from CAD to CAM allows for customized design and minimization of building components.

The ability to fabricate larger components minimizes erection time, as demonstrated by Boeing's decision to proceed with composite materials for its new 7E7 intercontinental passenger aircraft. Boeing studies have indicated that larger composite panels will lead to a drastic time reduction in the final assembly (Ponticel, 2003).

On the MMU, each panel geometry is exported from the parametric model to the variable form mold, or contour table, which sets the net shape mold for the lay-up of the composite panel. First, the core is draped upon the mold, and then the reinforcement fabric is "wet-out" (thoroughly soaked with resin), and placed upon the core. The resin is cured, thus freezing the shape so the panel can be removed from the mold. The panel is then completed by repeating the fabric lay-up on the opposite side. The finished panels are assembled into the designed structure and all joints are sealed and secured with additional fabric, creating a weatherproof monolithic shell.

The contour table

A digital version of Renzo Piano's 1967 conceptual pantographic variable form molding device, the contour table is an array of actuated pistons that are positioned according to a computer model (Figure 16.4). This digital model includes both the designed surface geometry (panel) and the planar surface of the contour table that controls the grid of pistons. The dimensions between the grid intersections and the surface geometry are output to servo motors on the table that locate the Z-axis of each piston, effectively approximating the desired topology (Figure 16.5). The MMU would contain a number of contour tables that can be configured to accommodate panels of different sizes and proportions. Multiple tables can then be linked, forming a larger array of pistons acting in unison.

Panel boundaries

The design of the shell as a whole must not only accommodate the functions required by the program, but also ensure that the derivation of the constituent panels reduces the fabrication and erection time. By using the minimal number of panels necessary to create the desired geometry, the number of time-consuming joints is reduced. In addition, the individual panels can be freestanding in order to expedite the erection process (Figure 16.6).

There are multiple ways to design and fabricate robust composite sandwich joints that are stronger than the sandwich itself. The preferred method is a type of scarfed joint (Figure 16.7), which not only provides optimal load transfer, but also seals the vulnerable core from environmental exposure and establishes a support to ease erection.

Applications

With the onset of contemporary freeform architecture, the need for efficient fabrication of complex-shaped cladding has continually grown. The MMU is well suited to such a task, providing a solution that could potentially integrate MEP systems into the cladding.

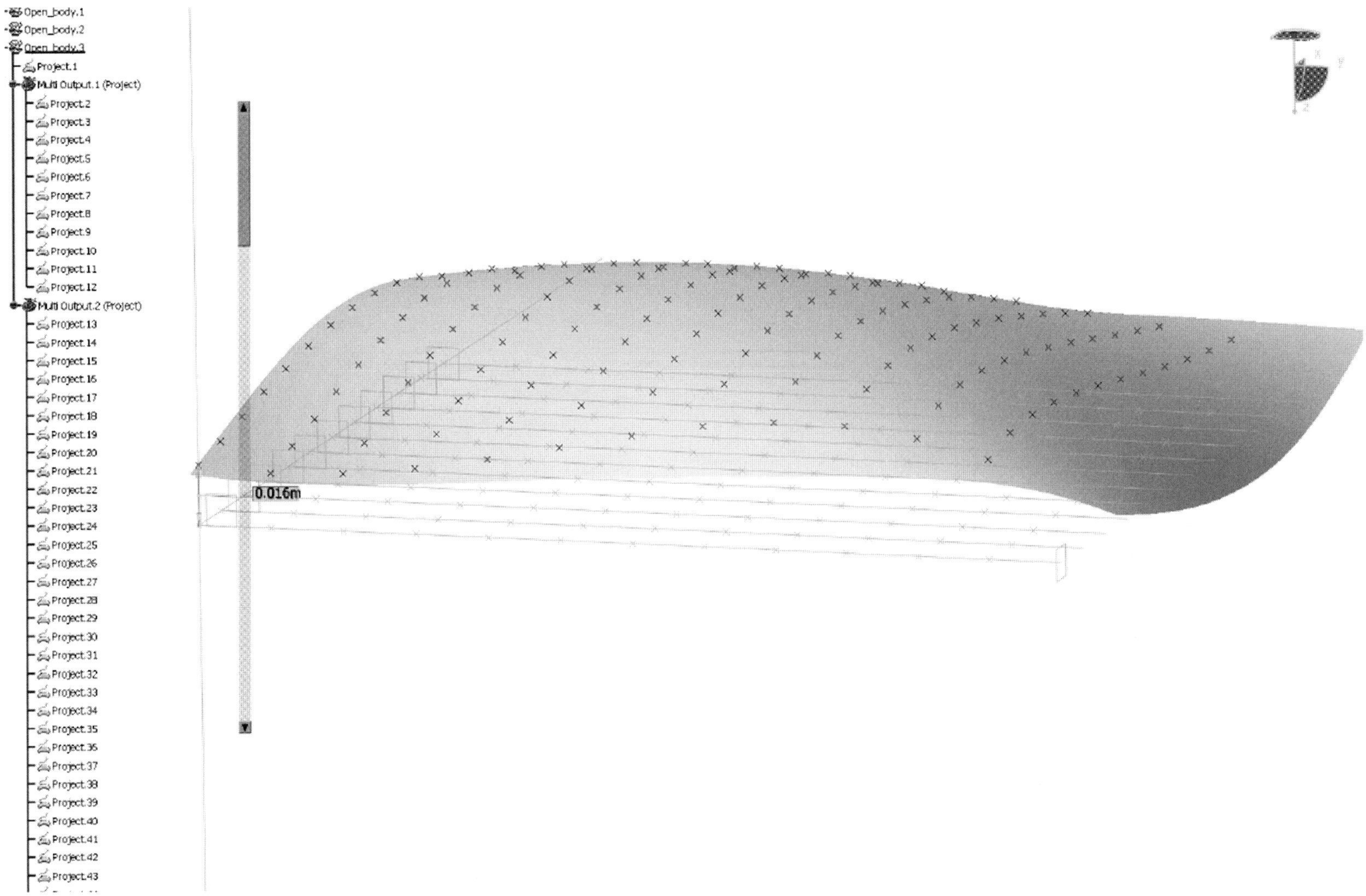

16.4 Parametric model of surface and projected grid points

16.5 First contour table prototype with cured carbon/epoxy laminate; the dimples indicate the locations of the piston heads

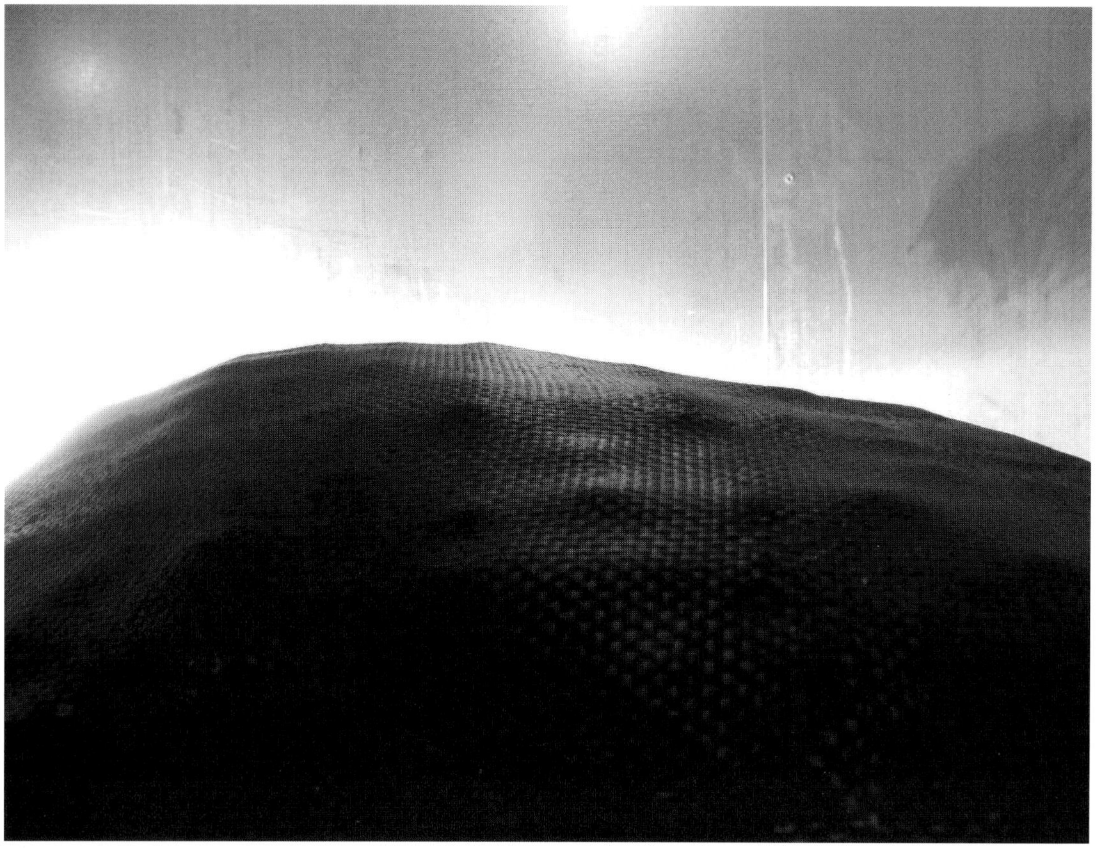

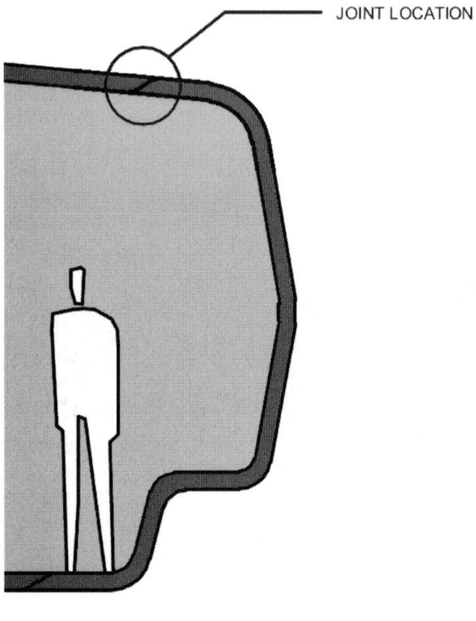

16.6 Example of possible joint locations

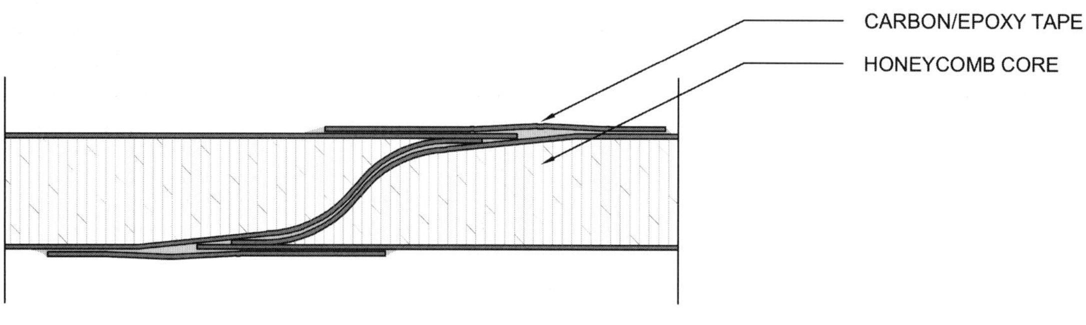

16.7 Scarfed joint

Freeform cladding systems are typically constructed as metal rib assemblies. Much engineering time is spent laying out the structural ribs and designing a connection system that can easily attach to the primary structure and accommodate shear forces while providing adjustment in three principal directions. The need to dimensionally adjust the connections typically arises from a disparity of tolerances allowed the different building trades. The concrete and steel contractors, who are still developing means to construct amorphous forms, are allowed much larger tolerances than external cladding contractors, given that water penetration and esthetic considerations do not allow for large gaps.

The MMU, coupled with a real-time coordinate system, could manufacture cladding on-site, providing panels that are properly adjusted to the actual connection points, eliminating the need to predict and accommodate potential dimensional variation with complex connections. It is, in a way, a "design-as-you-build concept." The real-time coordinates update the parametric model that directly feeds the manufacturing and informs the erection.

In addition to cladding, the MMU technology could be implemented to produce entire buildings. Preliminary studies have shown the complexities involved in introducing composites as primary structural members into the hegemony of contemporary construction in terms of economy, feasibility, and legality. There are no specific references to polymer matrix composite structural members in either American or European building codes (Reux, 2004).

Conclusion

The results of our experiments demonstrated that some of our original assumptions were incorrect and we often failed to foresee the complexity of anticipated challenges. For example, we predicted the difficulty of achieving synclastic curvature with honeycomb due to its anisotropic properties, but we did not anticipate a problem with expanding the core nor did we predict the structural dilemma of crushing the cells when forming it. Conversely, we anticipated problems with curing the resin in absence of vacuum bagging and a properly controlled environment, yet the curing process wasn't an issue. From this we learned that while academic literature is a foundation from which to speculate, it cannot possibly predict all experimental outcomes in new conceptual territory. Valid research can only flourish through theory *and* practice.

Most importantly, the work to date has provided some very exciting potential research. Due to many difficulties with the honeycomb core, we decided to investigate alternatives more thoroughly. We discovered syntactic foams that have the properties required for mobilized manufacturing: lightness, formability, compressive resistance, and expandability. These foams are comprised of thermoplastic microspheres that contain hydrocarbon gas. When heated, the plastic shell becomes pliable and the gas expands uniformly, resulting in a rigid foam that is 100 times the volume of its original resin form. With a few modifications to the fabrication process, sandwich panels could feasibly be constructed with this new core material.

We have not discarded honeycomb as a potential core material. Research into the fabrication methods of honeycomb core has yielded new ideas that could potentially eliminate the need for a variable form mold. We believe that it is possible to design a custom core pattern based on existing manufacturing techniques. We learned that when a honeycomb pattern is curved in one direction the entire panel naturally becomes a saddle shape (Figure 16.8). By implementing an alternate, non-hexagonal and non-uniform pattern it may be possible to change the result from a saddle into a predetermined shape. These possibilities have reinforced our belief that architects and engineers can inspire new opportunities in the built environment by

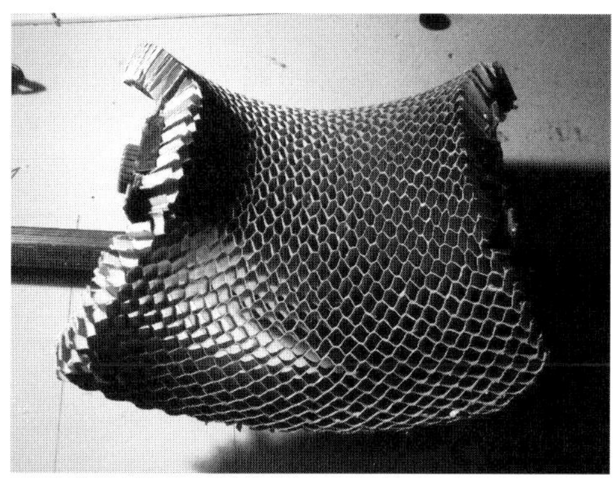

16.8 Anticlastic property of honeycomb

16.9 First full-size panel prototype: fiberglass, polyester resin, honeycomb

understanding the processes by which building materials are produced.

Currently, mobilized manufacturing is progressing along two fundamental paths specific to materials research and form-making. The implementation of the contour table has dictated further experimentation with syntactic foams and other formable core alternatives. Custom honeycomb cores could, however, be the future of the MMU and provide a higher degree of mobility than initially anticipated because the contour table would no longer be necessary. If this research succeeds in creating a material that takes its shape while being expanded, then Mobilized Manufacturing could truly be a lean process that synthesizes materials, fabrication, and erection into one seamless event on the building site (Figure 16.9).

References

Marshall, Andrew C. (1988) *Composite Basics-5*. Walnut Creek: Marshall Consulting.

O'Dell, Jane L. (1997) "Natural Fibers in Resin Transfer Molded Composites" in *4th International Conference on Wood Fiber-Plastic Composites*, pp. 280–285.

Ponticel, Patrick (2003) "Boeing Takes a Leap Forward with Composites" in *SAE Aerospace*, November, pp. 29–31.

Reux, Frédéric (2004) "Composites in Buildings: Promising Dynamics" in *JEC Composites*, pp. 7, 28–30.

Strong, A. Brent (1989) *Fundamentals of Composites Manufacturing: Materials, Methods and Applications*. Dearborn: Society of Manufacturing Engineers.

Transformable and Transportable Architecture with Scissor Structures

Carolina Rodriguez and John Chilton
University of Nottingham and University of Lincoln

The scissor concept

The scissor mechanism (Figure 17.1) is based on a very simple concept. Essentially, it consists of two rigid elements connected by a rotational hinge that allows them to move freely relative to each other about the hinge. The term 'scissor' was adopted fairly recently to describe this mechanism, due to its similarities with the cutting device. Multiple interconnected scissor units can form a structure that transports forces, and the resultant motion, from one element to the next. Hence, throughout the process of motion the entire system expands or contracts and thus changes in size and shape.

It is difficult to date the invention of the scissor mechanism precisely; however, history suggests some of its extensive and varied uses. There is evidence that the ancient Egyptians employed it in elementary folding chairs and ancient Mongolian communities constructed their portable yurts using grids of straight wooden bars linked by scissor hinges, a practice that continues to this day. Renaissance inventors such as Leonardo Da Vinci applied it to their machines, mobile bridges and umbrellas. For centuries this mechanism has been closely associated with terms such as deployable, foldable, mobile, portable and transportable. However, it was not until the nineteenth century and the emergence of the industrial revolution that the widespread use of this mechanism became commercially popular through its domestic and industrial use in, for instance, lamps, chairs and awnings.

In the architectural field, these kinds of systems only started to receive further academic attention in the second half of the twentieth century, encouraged by revolutionary movements such as Archigram. Portable architecture also began to play an increasingly important role in modern life style. Architects and engineers demonstrated a growing interest in studying and experimenting with these mechanisms since they offer important advantages over other systems because of the relative simplicity of their stress-free assemblies.

One of the first architects interested in this topic was Richard Buckminster Fuller (1895–1981), who reintroduced the concept of 'deployability' in structures as part of his synergy theory. He used a geometric model called 'Jitterbug' to explain the essence of deployable systems and illustrate how the loads behave structurally during the movement. Applying this concept, he designed the first known unfolding geodesic dome for the United States Army (Edmondson, 1987). One of Fuller's students, Emilio Perez-Piñero (1936–72), began work on this topic in the early 1960s (Candela *et al*., 1993). He became a pioneer in the construction of automatic deployable domes and different types of grids formed from straight scissors. Furthermore, Perez-Piñero proposed new deployable mechanisms with two layers of scissors and designed cylindrical shelters using bars with offset pivot (Figure 17.1). During his life he carried out a series of projects and won several prizes for his work. Recognized contemporary researchers and architects, such as Santiago Calatrava (Tzonis, 1999) have been inspired by Perez-Piñero's inventions. Calatrava made his own

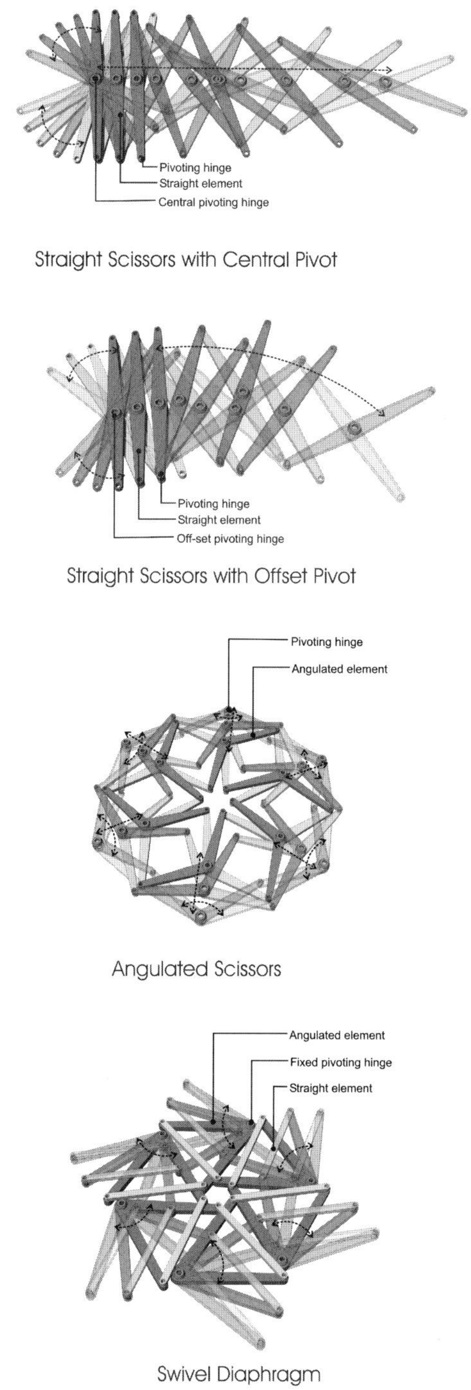

17.1 Deployable mechanisms

contribution with his doctoral thesis on the flexibility of three-dimensional structures, which he presented at the Swiss Federal Institute of Technology (ETH) in Zurich in 1981.

The stability of the structure during deployment has been one of the key issues of study. Large-size projects containing a considerable number of bars and hinges, subjected to long-term use, needed special study to ensure a proper and safe behaviour of the structure. Subsequent researchers led ambitious investigations into structural and loading aspects affecting these types of mechanisms. Felix Escrig (1996) and Juan Perez Valcarcel (Candela *et al*., 1993) developed innovative methods to solve problems of angular instability presented in large-scale structures with deployable meshes of scissors. Theodore Zeigler (1997) focused his investigations on the restrictions of movement existing during the process of deployment of spherical domes. Using additional members he attempted to avoid the self-locking phenomenon in the deployable assemblages. Charis Gantes (2000) detected incompatibilities between the members of certain scissor chains that produce locking in the structures and consequent stress. Subsequently, he formulated a theoretical method for obtaining geometric constraints for deployable units, maintaining the desired features of deployability and stress-free conditions.

Searching for more efficient and convenient geometry for foldable scissor structures, W. Shan (1993) applied Formex algebra to manipulate the configuration processing for pantograph structures. In a parallel study, K. Kawaguchi, Y. Hangai and K. Nabana (Kawaguchi *et al*., 1993) proposed an analytical procedure to obtain the quickest way to erect a scissor structure and the optimum processes to fold it into a desired shape. Faris Albermani and Travis Langbecker (Langbecker, 1999) used kinematics and non-linear analysis to determine the structural response of foldable structures in the fully deployed configuration under static loading. There was also a project developed by the Kinetic Design Group at the Massachusetts Institute of Technology (Block, 2003) looking at the advantages offered by scissor structures combined with structural membranes. Other researchers concentrated instead on examining adaptations to the original concept of scissor elements, in order to improve the system or the construction process. For instance, Zhong You and Yan Chen (Chen and You, 2001) made use of *three-dimensional* over-constrained linkages instead of *two-dimensional* foldable scissor hinges. Kokawa Tsutomo and Hokkaido Tokai (1997) included supplementary cables in scissor arches to make deployment easier.

Most of the projects led by the researchers mentioned above have concentrated on the engineering aspects of the mechanism and some limited exploration of its ability to change in size, since this characteristic is of great use for portable structures. Despite these efforts, little attention has been paid to a special, more architectural, attribute of the mechanism, which is its ability to change in shape and propagate movement.

Chuck Hoberman took a different approach in the use of scissor mechanisms. In 1991 he founded *The Hoberman Association*, a company specialized in the design of diverse artefacts based on the scissor mechanism. Amongst his products, the folding toys in particular have been very popular and sold successfully worldwide. Some of these toys use a new type of mechanism invented by Hoberman named the 'angulated scissor'. This new system is an alteration of the original scissor idea that consists of bending the rigid elements in order to achieve a desired angle and form polygonal shapes that are not possible with straight scissors. Employing this system, he has designed complex and exciting structures such as the Expanding Geodesic Sphere, Expanding Hypar, Iris Dome and Hoberman Arch. Further research on angulated scissors, with particular emphasis on applications within the aerospace industry and sports venues, has been carried out by Sergio Pellegrino, P. E. Kassabian and Zhong You (Kassabian *et al.*, 1997) and Frank Jensen (2001) at the Deployable Structures Laboratory, University of Cambridge, UK.

A recent alternative system based on the scissor mechanism and named the 'swivel diaphragm' has been developed by the authors at the University of Nottingham (Rodriguez and Chilton, 2003). In this system, angulated rigid elements are connected by straight bars forming a closed circuit, which pivots around external supports. The swivel diaphragm achieves a similar retraction to that of a ring with angulated scissors but additionally it provides fixed outer supports within the ring configuration. Such an attribute helps to simplify the complicated support conditions associated with the use of angulated scissors.

Angulated scissors and the swivel diaphragm offer fewer properties of compactability than normal straight scissors. Hence their potential for transportable architecture is more limited. However, the natural ability of the mechanism to modify its shape by a coordinated movement is highly valued within these systems, implying a completely different dimension that has not been fully explored: 'transformation'. This characteristic could be exploited, particularly within dynamic or kinetic structures, a field that has grown in importance in recent decades. This type of application could contribute to the development of a new trend in the use of the scissor mechanisms within 'transformable architecture'.

It could be argued that throughout history the employment of scissor mechanisms in architecture has evolved from a purely functional and practical perspective through transportable architecture towards a more artistic approach within transformable architecture. Even though the fundamental principle of the scissor mechanism still remains the same, two different functions and two almost extrapolated contexts could start to diverge. It is left to the designer's capacity to comprehend their variance and respond with a distinct architectonic strategy for each case. This exercise could prove complex because, despite the two typologies being of distinct form, they are not absolute. Certain architectural and structural aspects still permeate from one to the other. In order to help establish the division line between the two ideas, the differences between one category and the other, in terms of structural requirements, architectural aspects and potential applications, are discussed below. You will find in this discussion a different perspective regarding the key architectural factors that affect buildings with scissor structures and a more subjective analysis of their structural and morphological characteristics.

Architectural differences and potential applications

The major differences between transportable and transformable buildings are easier to identify in terms of their architectural aspects rather than their structural and morphological requirements. This is because between these two categories the function of the building is almost contradictory whilst the structural and formal aspects can be manipulated (in some measure) to serve one or the other. In terms of its nature, transportable architecture is absolute, linear and imposing; its character is more related to a mechanical universe. Transformable architecture, in contrast, is dynamic and evolving and exhibits the space-time-dependent qualities of an organic universe. In terms of applicability, there are two factors that help to differentiate transportable buildings from transformable buildings: the location (where the structure is placed) and the time-frequency for which the system is used. A regular change of location implies that the structure should be easy to pack and transport, rapid to erect, simple to dismantle and ready to be reused afterwards. Primary examples of this kind of temporary and transportable buildings are: itinerant entertainment venues, marketing stands, military installations, emergency shelters, short-term constructions for health, education and sports activities.

An experimental project for a travelling theatre, proposed by the authors (Figures 17.2, 17.3, 17.4), illustrates a potential application of scissor structures in transportable architecture. The enclosure of the venue consists of a lightweight network of straight scissors covered by a flexible membrane. The trucks that transport the folded assembly are an integral part of the design. The sides of the truck fold back to form a raised floor for the venue. The structure is gradually deployed in two directions assisted by a hydraulic platform-lift attached to the back of the truck. When the structure is fully unfolded and secure it becomes completely self-sufficient as regards anchorage. This enables the building to be erected practically anywhere.

Transformable structures, in contrast, are generally conceived to operate at a permanent location responding to a specific time-frequency cycle. Buildings that need to be able to adapt to diverse changing conditions suggest the use of a transformable structure. Examples are buildings specifically designed to respond to variable climates, to accommodate different functions, to extend or change in size or to transform their aesthetic features and architectonic scenarios. These changing conditions are the key factors that help to define the cycle of movement of the structure. For instance, if the structure is designed to respond to different weather conditions the scissor mechanism used should be able to synchronize with the fluctuations of the climate. To some extent, these kinds of buildings act as living organisms, sensitive and responsive to the environment that surrounds them. The types of uses mentioned above are relatively new in the architectural sphere compared to those of transportable buildings.

One of the most likely applications of scissor structures for transformable buildings is in retractable roofs, mainly for sports venues. Currently, there is an increasing demand in the sports business for retractable roofs in an attempt to increase the flexibility of the venues and at the same time maximize revenues. So far there has not been a large-scale retractable roof built employing scissor mechanisms, but research in this area suggests that it is only a matter of time before this could become a reality. Figures 17.5–17.7 show a design for a possible retractable roof using a dodecagonal swivel diaphragm ring. The retractable roof is constituted by triangular rigid frames that hold up inflatable cushions that form the cover. The frames pivot in a coordinated manner from fixed external supports helped by two layers of straight bars that join them together. The covers meet at the centre of the ring in the fully closed position and deploy towards the perimeter at the fully open stage.

In addition to retractable roofs, scissor mechanisms have the potential for adaptability that could be extensively exploited from the ecological and sustainable point of view. It is widely known that since the early 1970s there has been a growing awareness and concern about the negative environmental effects of the increase of energy consumption for heating, cooling and internal climatic control within buildings. For this purpose, devices with these types of systems could be designed to control day lighting, ventilation and/or sound insulation. The experimental project 'transformable skin' shown in Figures 17.8–17.9 proposed an alternative solution for shading control. It uses a grid of interconnected swivel diaphragms that, in conjunction with photocells, deploys in order to regulate internal lighting conditions. A similar concept for shading devices was applied by Jean Nouvel in 1988 in a design for the main façade of the Institute of the Arab World in Paris (though the mechanism used in that case was different as each module had to be deployed individually).

17.2 Travelling theatre, deployment process

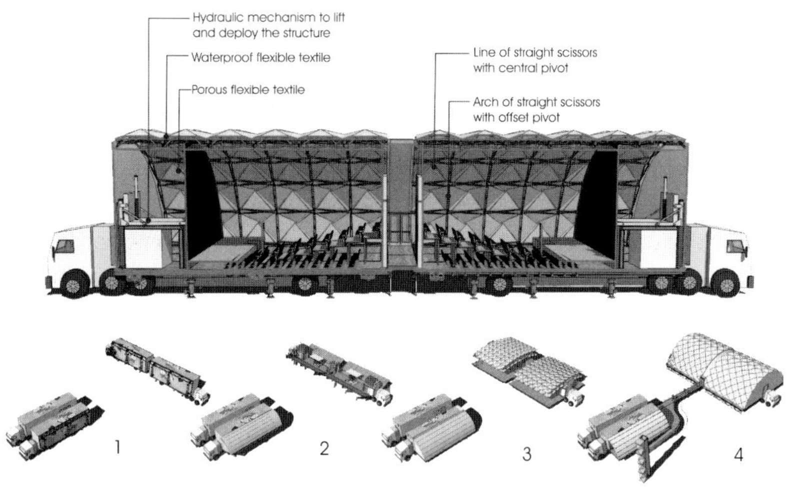

17.3 Travelling theatre, deployment process

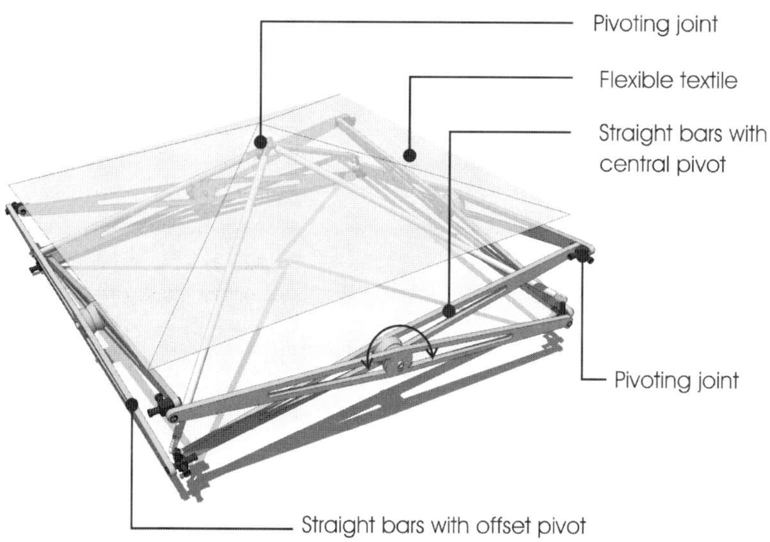

17.4 Travelling theatre, scissor module

17.5 Retractable roof

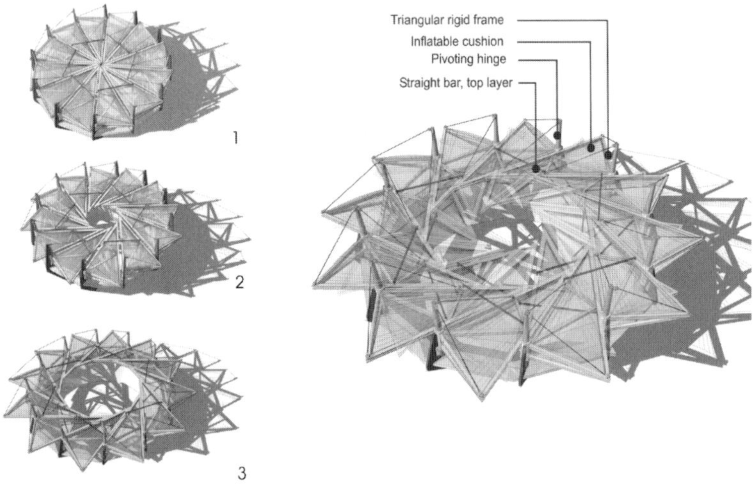

17.6 Retractable roof, deployment

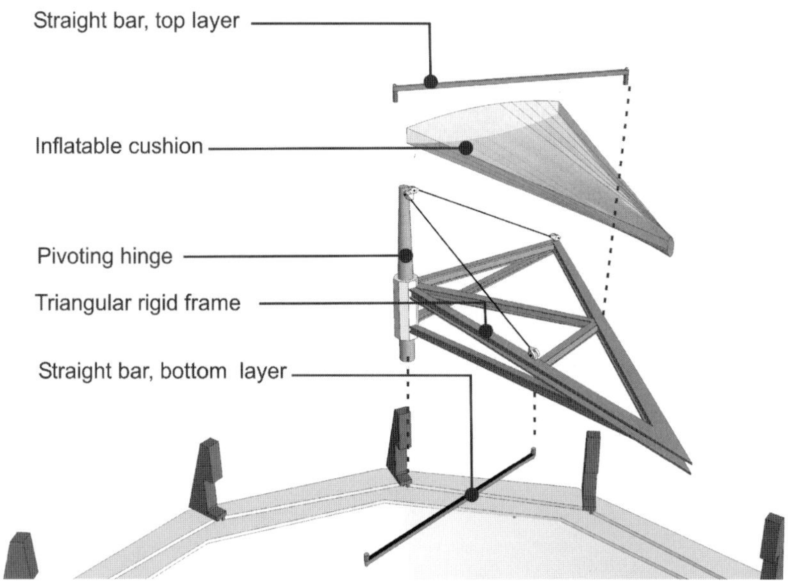

17.7 Retractable roof, cover module

17.8 Transformable skin

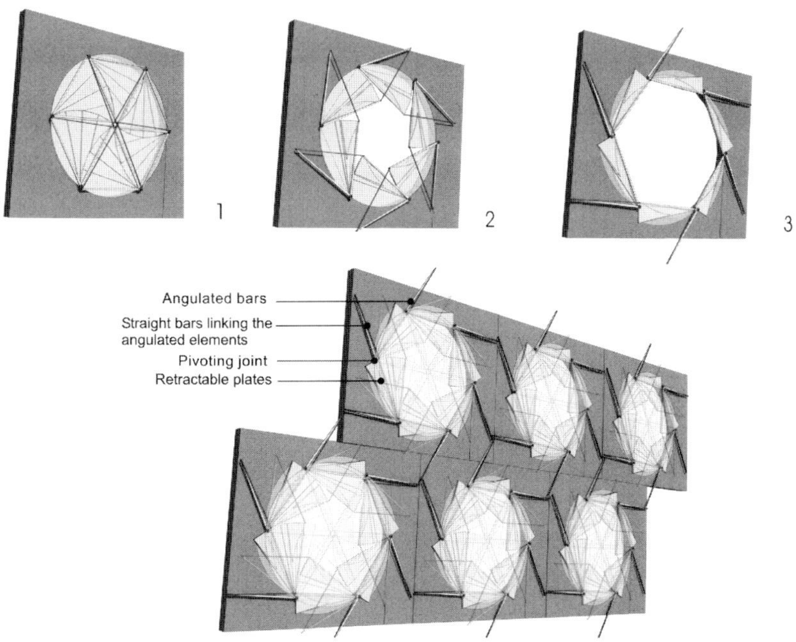

17.9 Transformable skin

Another potential use of these mechanisms in transformable architecture is in canopies. With the invention of new scissor mechanisms the original umbrella, which initially employed straight scissors, could be adapted to serve other purposes. Umbrellas are not necessarily portable any more; an example of this are the 12-metre-span convertible umbrellas designed by S. L. Rasch (Otto and Rasch, 1995) to cover the courtyards of the Holy Mosque in Medinah. In summer, these umbrellas are open during the day to shade the courtyard whilst they are closed at night to promote cooling by radiation to the night sky. Conversely, in winter, the umbrellas are closed during the day to allow the sun's rays to warm the courtyard whilst at night they are opened to trap warm air and limit radiation to the cold night sky. Other alternative canopies might be proposed with the intention of being purely transformable such as the example illustrated in Figures 17.10–17.11. This idea for a canopy, proposed by the authors, compromises a hexagonal swivel diaphragm ring formed by triangular rigid frames that support translucent and lightweight plastic covers. The covers overlap at the closed stage and retract at the open stage, increasing the area of coverage by more than 50%. Additional membranes attached in between the frames retract to form a hexagonal shape when the canopy is fully open.

Only a limited number of architectural projects using scissors mechanisms have been built to date. This scarcity of built examples suggests that there are various factors, still not clearly specified, restricting the development of practical applications. Possible reasons for this are that the systems are relatively new, their capabilities are not well understood and/or specialized literature on the subject is not widely available. Their complexity and relatively high cost may also have some influence. The authors have identified three architectural factors that, if taken into account when designing, might help to exploit the capabilities of scissor mechanisms to the limit. They are: the aesthetics of the movement depicted by these buildings, the coexistence with the environment that they inhabit and their interface with the visitor and observer.

Aesthetics of the movement

The process of movement of a transportable structure with scissor mechanisms is in one sense linear. It starts in its folded state and progresses to its fully unfolded position or vice versa. However, the displacements produced by angular rotation of the scissors are generally non-linear. It is highly probable that the observer will, for most of the time, appreciate the architecture of the building in one of these two states. Hence, more attention needs to be paid to the aesthetic image that they reflect when packed and being transported and when unpacked and functioning. It is paramount that a packed structure expresses in some way its lightness, manoeuvrability and furthermore transportability. When it is fully unfolded, the structure should transmit the idea of strength and give an impression of security and resemble a final product instead of an assembly in a stage of deployment.

The process of movement of a kinetic or transformable structure is completely different. It has a metamorphic nature that allows it to evolve from one stage of movement to the next, each being of equal importance in terms of aesthetics. Within the capabilities of the scissor mechanism it follows rhythmical kinetic cycles with no specific initial or final positions. The motion is gradual but not necessarily sequential. Thus, the system could be coordinated to a completely chaotic pattern and be capable of responding to unpredictable stimuli such as weather conditions. Certainly, the structure will disclose a different appearance at every stage of the motion and it will also affect the overall architecture of the building in a distinct way during this process. Such an attribute could become a very powerful tool of design, making it possible to achieve different

17.10 Transformable canopy

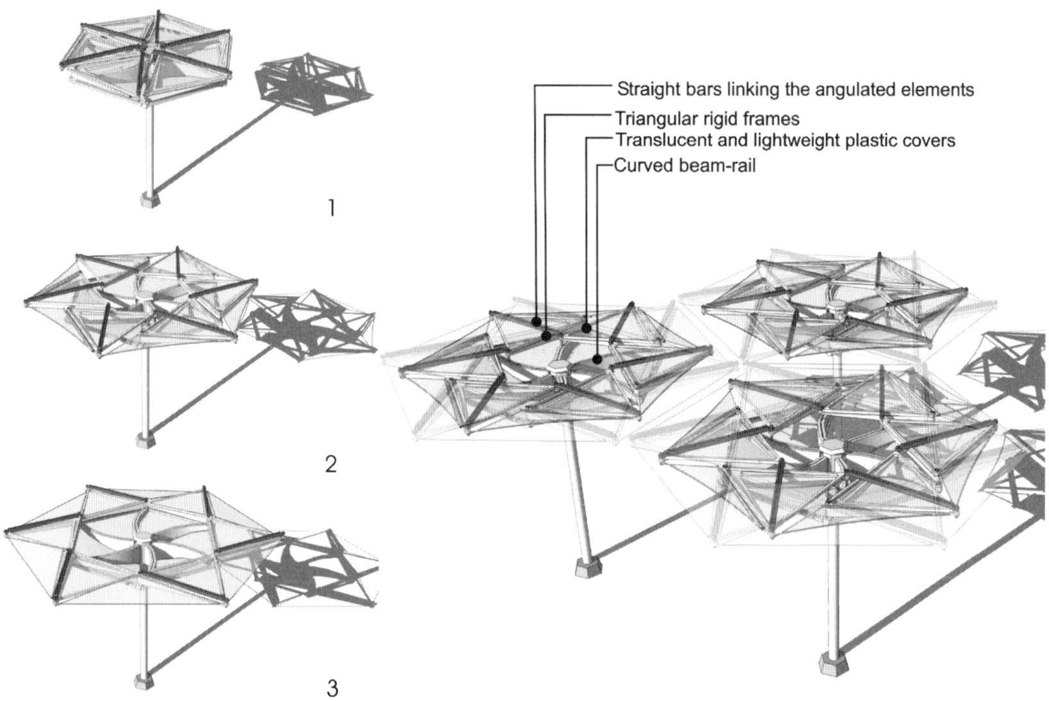

17.11 Transformable canopy, deployment process

manifestations throughout the 'performance' of the structure by working with the shape and geometry of the scissor elements.

Coexistence with the environment

Transportable buildings with scissor mechanisms are of a timeless and delocalized nature. Their relationship with the environment is relatively elusive – to a certain degree they have the flexibility to adapt to different locations but at the same time they are not able to respond to any particular landscape. They could be defined as temporary visitors that impose on the environment for a limited period but rapidly move away, leaving little or no trace of their passing. Their performance and duration are determined mainly by their function rather than their surroundings. That is partly the nature of this category; however, this could also make them unsuitable for locations such as historic and conservation areas or nature reserves where the surroundings are extremely sensitive to alien constructions. To address the negative impact of the transportable structure it could be designed to be autonomous and architectonically conscious of the diversity of environments.

Alternatively, the way in which transformable buildings interact with the local environment is completely opposite. Due to the fact that they are fixed in a permanent location, they need to adopt a more considerate position towards the landscape attributes within which they are built. This kind of architecture should not only coexist but also cooperate in synchronicity with the landscape. The buildings should be designed in order to comprehend and respond to environmental changing conditions as a smart and 'living' entity.

Interface with the observer

Transportable architecture should be envisaged as ephemeral; one day it is there, the next it is gone. Due to its temporary existence, the observer can watch the process of making, using and dismantling architecture in a relatively short time. This type of architecture has the magical capability to emerge from nothing, serve a purpose and then disappear. Consequently, the existence of the building relies on memory and it is essential to make a strong impact in order to survive in the observer's thoughts.

Transformable architecture often has more of a long-term character. It diminishes the importance of the initial impression and emphasizes the importance of the permanent integration with the observer. The social acceptance, comfort and response of the people who inhabit and interact with this constantly changing architecture are of great priority for its accomplishment. In fact, the argument of the practicability of these structures and their future use relies on the level of acceptance and how well they fulfil the observers' or inhabitants' expectations. They must go beyond the stage of novelty and become part of a familiar space in daily use in order to be successful as permanent transformable architecture.

Structural morphology

The morphology of the scissor structure is a great influence on the success of its ultimate performance. The shape of the scissor components, their scale, modularity, frequency and material are crucial variables that can be manipulated in different ways to define the transformable or transportable nature of the building. These aspects of design are examined here from an abstract point of view without involving specific considerations of particular projects or complex calculations.

Structures with various interlinked units of scissors form elaborate grid configurations that adopt a synergetic nature: Synergetic because analysing the character of a single scissor component is not enough to establish the operation of the grid as a whole. This can only be understood by studying the relationship between all the units and the way they are interconnected. Even when the assemblages are modular, the structural study of scissor frames cannot rely on the analysis of isolated scissor units; it has to comprehend the structure as an absolute entity.

Type of components

The first aspect to identify when designing a scissor structure is precisely the type of scissor and connection to be used according to the function. The strut-angle formed between the pivoted hinges in the scissor unit is the chief factor that determines the morphology of the overall structure. It also dictates the folding direction of the frame and its final degree of deployment. Most of the existing types of scissors could be classified according to the value of the strut-angle and its position along the rigid element into:

- straight scissors with central pivot
- straight scissors with offset pivot
- angulated scissor
- multi-angulated scissors
- and the swivel diaphragm

These are all variations on the basic mechanism but each of them displays a distinct character and a different potential for use. Assemblages of straight scissor bars, with a strut-angle of 180° and central pivot allow the structure to compact into very small packages when fully folded and to form an almost flat surface when fully unfolded. Similarly, structures using straight scissors with an offset pivot can be folded in a fairly compact bunch, the difference being that they enable curvatures to be formed within the structure. Thanks to such characteristics, designs employing the two types of scissors mentioned above represent ideal options for transportable buildings. On the other hand, configurations where the strut-angle is smaller than 180°, such as with angulated scissors, multi-angulated scissors or the swivel diaphragm, do not vary greatly in shape during the deployment process. Regularly, the dimensions change in proportion to the magnitude of the strut-angle (the smaller the angle the smaller the change). However, it can be argued that generally, structures using angulated scissors or swivel diaphragm do not have an exactly defined folded and unfolded position. Instead, they experience sequential stages of 'mutation', all equally valuable, which establishes them as attractive and innovative alternatives for transformable architecture.

External supports

Ideally, structures designed for transportable buildings should be stress-free assemblies that solve their loads internally without the need for special external supports fixed to the site. This enables the building to be erected practically anywhere. With that purpose, various solutions have been developed to restrain the deployment of the scissor units and self-lock the structure in the unfolded state. For instance, flexible elements such as cables or wires, strategically attached to the ends of the specific scissor units, constitute a practical and easy alternative. They become tensioned by the structure when deploying, thereby limiting the movement. On the contrary, in transformable constructions the forces do not necessarily have to be solved internally. The structure can make use of external supports or be fixed permanently to the site. Solutions for external supports may be relatively complex since they transmit load from a kinetic frame to a static foundation. Rails are a straightforward option for external supports for some transformable structures but they require constant maintenance to avoid complications due to friction. Pivoted supports are more efficient since they minimize the contact area between the mobile structure and the support.

Material, scale, modularity and frequency

Lightness and strength are two of the major requirements of scissor systems. Therefore, the material chosen to build the scissor structure has to fulfil these needs. Standard rigid materials such as wood, metal, plastics and composite materials are currently used. The material of the joints should offer high levels of fatigue resistance and at the same time allow a smooth pivoting hinge avoiding excessive friction. To minimize joint loads, loads in general and, therefore, element section sizes, cladding materials should also be as lightweight as possible (e.g. textile membranes).

As in many types of structural system the demands magnify as the scale increases and for structures with scissors this is a very important issue, as their transportation, erection and motion become more complex. Scissor structures of small to medium scale can be designed to operate manually or through a simple mechanical procedure. Those of larger scale demand a higher technological investment in the design of the structural components and the movement process to ensure their efficient operation. As well as the scale, the number of units in the overall structure is a key factor in determining the performance of the complete scissor structure and the anatomy of its components. For instance, because of the larger number of joints, structures with a high frequency of scissors are likely to suffer more problems during deployment and require more maintenance.

Due to the fact that scissor structures consist of numerous identical elements, standardization and modularity are of high priority in order to lower production and assembly costs. Modular components allow more flexibility in the design, since complete scissor units can be added or subtracted to explore different configurations. All the aspects mentioned above are interrelated and interdependent and these factors have to be balanced according to the function and specific needs of a particular project in order to achieve the desired character for the building.

Conclusions

The study of the implications of the use of scissor mechanisms in architecture is still in its infancy, especially with regard to their real practicability and user-friendliness. Knowledge in this area can really only be achieved by trial and error. This represents a great challenge for architects, engineers and researchers who are in the business of convincing clients of the employment of scissor structures. There is a relatively extensive stock of knowledge concerning the geometrical and structural aspects of scissor systems, accumulated through years of theoretical research. Having this scientific research as a platform certainly helps to support the investigation of potential architectural applications. A multidisciplinary investigation of the effects that architecture with scissor mechanisms produces in those who regularly experience them is urgently required. This is particularly so as there is an emerging trend in the use of these structures within transformable buildings. Despite the technological advances in this area, the human perception of architectural structures in continual motion is a topic still unexplored. The ongoing debate surrounding the real potential of scissor structures has become even more complicated with their introduction into the complex sphere of transformable architecture. Are we prepared to inhabit buildings that constantly change their architecture? What type of sensations can we experience within these mutating spaces? Further analysis of the impact caused by employing these complex systems within permanently habitable buildings is vital to direct the future development of these structures towards a gentle integration with the architectural environment and, in this way, to obtain users' acceptance.

References

Block, P. (2003) 'Interactive Kinetic Structures: Architecture with an Organic Trait', 4.221 Joint SMArchS Colloquium, BT Seminars Series, Massachusetts Institute of Technology.

Candela, F., Perez, P. E., Calatrava, S., Escrig, F. and Perez, V. J. (1993) *Arquitectura Transformable*, Seville: Escuela Superior de Arquitectura de Sevilla, Spain.

Chen, Y. and You, Z. (2001) 'Deployable Structural Elements Based on Bennett Linkages' American Society of Mechanical Engineers International Congress, 11–15 November, New York, USA.

Chilton, J. C., Choo, B. S. and Wilkinson, D. (1998) 'A Parametric Analysis of the Geometry of Retractable Reciprocal Frame Structures' *Lightweight Structures In Architecture, Engineering And Construction*, Vol. 1, 547–555.

Deployable Structures Laboratory Web Page: http://www-civ.eng.cam.ac.uk/dsl/.html (accessed 10 October 2003).

Edmondson, A. (1987) *Fuller Explanation: The Synergetic Geometry of R. Buckminster Fuller*. Boston: Birkhäuser.

Escrig, F. (1996) 'General Survey of Deployability in Architecture', *Mobile and Rapidly Assembled Structures II*, Computational Mechanics Publications, 3–24.

Gantes, C. (2000) *Deployable Structures Analysis and Design*, Billerica, Mass.: Wit Press.

Hernandez, C. and Zalewski, W. (1993) 'Expandable Structure for the Venezuelan Pavilion at Expo '92' *Space Structures 4*, eds G. A. R. Parke and C. M. Howard. London: Thomas Telford. Vol. 2, 1710–1719.

Hoberman, C. Web Page: Url>www.hoberman.com (accessed 1 October 2003).

Hoberman, C. (1996) 'Temporary Unfolding Structures', *Detail; Temporary Structures*, Munich.

Jensen, F. V. (2001) 'Cover Elements for Retractable Roof Structures', First Year Report, Ph.D. University of Cambridge, UK.

Kassabian, P., You, Z. and Pellegrino, S. (1997) 'Retractable Structures Based on Multi-Angulated Elements' in *Proceedings of the International Colloquium Structural Morphology*, ed. J. C. Chilton *et al.* University of Nottingham, 92–99.

Kawaguchi, K., Hangai, Y. and Nabana, K. (1993) 'Numerical Analysis for Folding Space Structures' *Space Structures 4,* ed. G. A. R. Parke and V. M. Howard. London: Thomas Telford. Vol. 1, 813–823.

Kinetic Design Group Web Page: http://kdg.mit.edu/ (accessed 1 November 2003).

Langbecker, T. (1999) 'Kinematic Analysis of Deployable, Scissor Structures', *International Journal of Space Structures*, 14(1).

Otto, F. and Rasch, B. (1995) *Finding Form, Towards an Architecture of the Minimal*, Catalogue for the Exhibition in the Villa Stuck, Munich, Germany.

Popovic, O., Chilton, J. C. and Choo, B. S. (1997) 'The Variety of Reciprocal Frame (RF) Morphologies Developed for a Medium Span Assembly Building-Case Study', *Structural Morphology: Towards the New Millennium,* ed. J. C. Chilton *et al.* University of Nottingham, 164–171.

Rodriguez, C. (2000) *Arquitectura Metamórfica*, Bogota: ICFES and National University of Colombia.

Rodriguez, C. and Chilton, J. (2003) 'Swivel Diaphragm: a New Alternative for Retractable Ring Structures' *Journal of the International Association for Shell and Spatial Structures,* 44, 181–188.

Shan, W. (1993) 'Configuration Studies of Foldable Structures' in *Space Structures 4*, eds. G. A. R. Parke and C. M. Howard, Thomas Telford, London, Vol 1, 824–832.

Tsutomu, K. and Tokai, H. (1997) 'Cable Scissors Arch-Marionettic Structure', *Structural Morphology: Towards the New Millennium*, ed. J. C. Chilton *et al.* University of Nottingham, 107–114.

Tzonis, A. (1999) *Santiago Calatrava: The Poetics of Movement*, London: Thames and Hudson.

Wan Ho, M. (1997) 'The New Age of the Organism', *Architecture Design Profile No. 129. New science- new architecture*, 67 (9/10).

You, Z. Web page: URL>www-civils.eng.ox.ac.uk/~zy/research/publications.html (accessed 20 November 2001).

Zeigler, T. (1997) U.S. Patent. 4.026.313.

A Very Rapid Deployable Canopy System

Neil Burford and Christoph Gengnagel
University of Dundee and Technical University of Munich

Background

Arena Seating is the UK's leading supplier of temporary outdoor events seating. Their seating systems use a proprietary seat unit, which is supported by two separate types of modular under-structures. A proprietary canopy system comprising a simple aluminium portal frame structure is used on stands up to thirteen rows deep. The canopy system is cumbersome and slow to erect and is not really economically viable for general commercial use beyond the company's own requirements. Additionally, its erection procedure presents a number of clear dangers for the operatives involved.

This project was initiated when Arena became involved in developing a completely new grandstand system with the German manufacturers of the PERI-UP scaffold system to replace their original seating, which had been in service for twenty years. This gave Arena the opportunity to rethink its original canopy designs in parallel with the design of the new seating system. It was the company's intention to produce a new stand-alone seating and canopy system that would resolve the existing problems of their current systems and that would have significant export potential as a stand-alone, own-brand product.

The new canopy structure was required to provide a number of significant benefits:

- to be erected much more quickly and safely than existing canopy designs
- to have significantly reduced operational costs
- to bring added value to existing seating products by providing unique aesthetics and improved visual impact

These attributes would be particularly important at prestigious international events such as motor racing and golf, particularly in Europe and the USA.

The new canopy (Figure 18.1) is based on a concept first developed in 1996 within the Lightweight Structures Unit. The concept was for a deployable truss that uses a flexurally formed, bending-stiff member to facilitate its erection. It was not until July 2000, with the award of a Teaching Company Scheme grant, that a programme could begin to develop a canopy system based on the original concept. The project culminated in July 2002 with a full-size, functional prototype and the pre-production designs for the final canopy including its construction details and prototypes of principal connections.

The TCS Directorate provided joint funding to the University of Dundee, with a 50% contribution to the costs by Arena Seating for the research, prototyping and testing. This also funded the employment of two researchers – an architect and an engineer. In parallel, British Council/German Academic Exchange Service (DAAD) funding was acquired with the Technical University of Munich in order to determine the generic behavioural aspects of the structure. It was at this stage that the detailed design discussions were commenced with Arena Seating, to define the exact requirements for the new design.

18.1 Full concept model and PERI seating system, 2003

Introduction

This chapter describes the design and decision processes taken in the development of the prototype structure. There were three principal stages in the development of the canopy.

The first priority was to carry out a detailed evaluation of Arena's existing systems, procedures and manufacturing capabilities. Additionally, assessments were carried out for the structural requirements of the seating structure, the storage, handling and assembly requirements of the canopy, the environmental loading conditions and the associated cost implications for the design.

The second stage involved developing a full-size physical model to evaluate the overall behaviour of the structure both during and after deployment. This model was also used to determine the functionality of the individual components and to test prototypes and mock-ups of node joints and connections prior to manufacture. The information gained from building and testing the model was used to further refine the design based on a better understanding of the problems.

The last stage was to develop the principal elements and components of the system to improve their functionality and to optimize their design for manufacture. This involved using a number of visualization and testing techniques including sketches, simple mock-ups, computer-generated models, accurate stereolithography models and dimensionally and functionally accurate prototypes in the final materials of construction. The investigations focused on increasing durability, strength, structural performance, ease of use and integration between parts at the same time as reducing manufacturing and operational costs.

Existing seating and canopy structures

Arena's seating modules are based on a purpose-designed seating unit that is supported by two types of seating under-structure, namely: a purpose-designed A-frame system and an adaptation of the KWIK Form scaffold and form work system. The initial investigations revealed that the A-frame system was too light and would not be capable of withstanding the additional loads it would need to carry as a result of the new canopy. Consequently, the design focused on optimizing the system for use with the KWIK Form under-structure.

Seating system

The seating system is a purpose-designed galvanized steel structure consisting of a series of RHS beams connected laterally by either seating units or step units. The system has been in use for over twenty years and presents clear dangers for the operatives during its erection when they are working above head height. Currently the system can only be erected by inserting the seats first, before the deck. Modifications were proposed to adapt it to allow all the decking units to be inserted first with the seating units being inserted at a later time to allow the new roof system to be erected correctly and with greater safety.

Scaffold system – KWIK Form

This is ostensibly a modified scaffold system that is employed as an alternative to the A-frame. It is used predominantly on uneven sites where the A-frame system cannot cope. It is more time consuming during the initial stages of erection due to the need to attain a level scaffold base before the erection of the grandstand structure itself can commence. Once the base has been completed, operatives believe that it can be erected in times comparable with the A-frame system.

Existing canopy system

The canopy is a frame-type structure constructed from proprietary galvanized steel hollow sections. A 9-metres-wide by 13-metres-long bay comprises a roof formed by four principal trusses, which are connected perpendicularly by purlins. The roof is supported on vertical columns at the front and rear of the seating stand, which are rigidly connected to the trusses. The trusses and legs are assembled on the ground before being lifted into position by hand. The purlins are located and the membrane is pulled into position by operatives working at height, from ladders, making the erection process potentially dangerous.

Theoretical concept for a deployable truss

The aim of the project was to develop a new canopy system that could be easily, quickly and safely deployed. The principal objectives were that the new design should require a minimum number of operatives for the erection; it should not rely on large mechanical plant to lift elements of the structure and it should be capable of being erected from the ground.

The original theoretical concept was to work with a long, thin, initially straight member that had minimal bending resistance. Due to the relative proportions of its length and depth the member will be very flexible. If it is held at one end and subjected to a force along its axis and a small displacement force perpendicular to its axis, the member will deflect into a new curved form. A simple demonstration of this effect can be obtained by taking a horizontal member, pinned at one end and resting on a second support containing a roller bearing a given distance along the member from the pin (Figure 18.2 (top)). By applying a small vertical force to the end of the member, it will bend about the second support point, thereby producing an arched cantilever.

A second vertical displacement of the entire member can be obtained if the horizontal member is supported and pinned at a point a short distance in from the end, producing a dual cantilever (Figure 18.2 (bottom)). If the end closest to the pin is subjected to a downward, vertical displacement the member initially bends downwards along its length on either side of the pin. If the downward force on the short side of the cantilever is greater than the self-weight and downward force on the long side of the cantilever, the resulting effect will be an upward displacement of the end of the cantilever furthest from the pin. The member will have assumed an overall curved geometry.

Half-scale truss model

This concept gave the basis for a deployable cantilever. However, in this simple form, the member alone has very little stiffness and is of no practical use. To provide stability from further deflections a truss was conceived, consisting of a number of vertical compression struts, pinned above the flexible chord and connected along the top by a flexible steel cable and between the top and bottom of the adjacent strut by diagonal cables (Figure 18.3). In its flat state the diagonal distance between the top of the strut and the bottom of the adjacent strut is shorter than in its deployed or curved state. Therefore the cable's flexibility allows the compression member to straighten and the truss to flatten. Prestressing the truss at the end using a vertical tie-down cable rigidizes the system.

A rudimentary half-scale model was prototyped in the workshop at Dundee to prove the concept. It comprised a primary compression chord that was constructed from flat sections of pultruded glass fibre, A-frames from tubular aluminium, flattened at their ends and connected to fabricated steel brackets with simple pinned connections. The cables were simply attached to the pins using flat bar and ferrules. It was obvious that the model

18.2 (top) Deflection of a cantilever under a single vertical force and (bottom) deflection of cantilever under a single vertical load and a displacement of the rear support

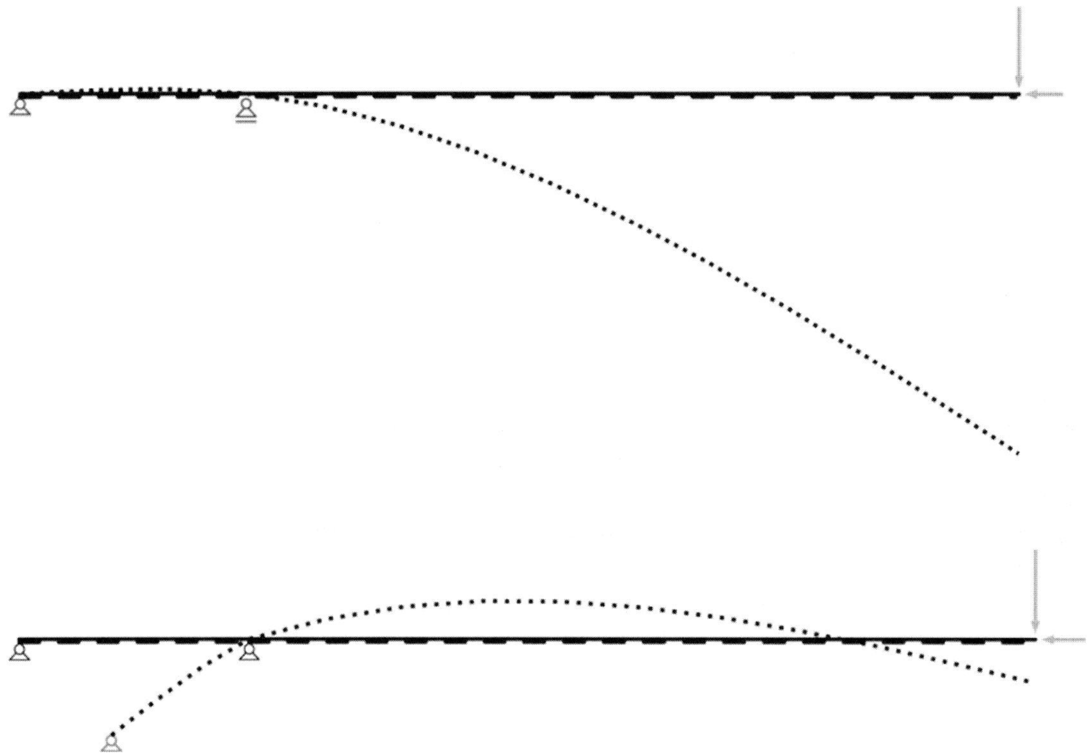

18.3 Erection sequences of the half-scale truss model

would require significant development in a number of areas. These included developing the three-dimensional form of the enclosure and its modularity, resolving the handling of the components and parts for transportation and deployment and developing a more comprehensive understanding of the loading and behavioural characteristics of the canopy.

Full-size concept model

A full-size concept model of the canopy system was produced that contained the basic functionality of all the key elements of the construction. It was the intention that this model would allow a more accurate assessment of the structure's behaviour during deployment and after deployment when subjected to applied loads. A particular concern was the frequency of the structure under certain wind conditions and therefore the weight, geometry and stress state of the system needed to be accurately representative of the final structure. It was also the intention that the model would allow the structure and components to be tested on the model prior to committing to a final design. This would be useful in order to refine and develop the criteria for the detail design of the components prior to committing a design to manufacture.

Form and general arrangement of the structure

The concept model (Figure 18.4) is a fully functioning, modular design comprising sectional aluminium trusses (A) with a 13-metre free span cantilever, supporting a stressed, anticlastic membrane skin (B), covering the seating (C). The trusses, spaced at 6-metre centres longitudinally, are located on a pivoting connection (D) at the rear of the grandstand and connected by tubular aluminium purlins (E) at approximately 3-metre centres along each truss. The canopy is anchored at the front by a steel cable (F), that also acts as the drainage system, connected between the membrane 'cone' and the seating support structure. Sway bracing cables (G) prevent the structure from racking laterally.

Each truss (Figure 18.5) consists of two purpose-designed aluminium extrusions (a.), which are approximately 150 mm wide by only 30 mm in depth. The flexibility of the compression chords over their 17-metre length allows the truss to be bent elastically into its working shape as it is erected. The truss is defined by a series of triangular A-frame assemblies (b.), which are attached, on pivotal connections (c.), to the top of the compression chords. The A-frames support the stainless steel top cable (d.) and shear diagonals (e.). They progressively reduce in height and width from the back towards the front, giving the truss its distinctive tapered form. The length of the top cable and the diagonals in the truss determine the resulting profile of the canopy and the truss action provided by these elements restrains the chords from further downward deflection due to imposed loads. The back diagonals (f.) are required in the first three bays of the truss.

Canopy erection process

From the outset the design of the canopy was driven by the need to improve the safety for the operatives during the assembly, deployment and dismantling of the canopy. A number of changes and developments were required from the original concept. The most significant of these was the requirement to sub-divide the main trusses into a number of separate sections in order to reduce the weight and bulk of the components during transportation. This meant that the assembly process needed to be rethought in relation to the seating platform and the design of the node joints (Figure 18.6).

Stage 1

Each truss is assembled from four separate sections on the seating platform. The rear section of the truss is first of all located on the pivot on the seating upright. An electric winch is connected to the end of this section and it is lifted to an appropriate height to allow the connection of next section of truss. This is repeated for each section.

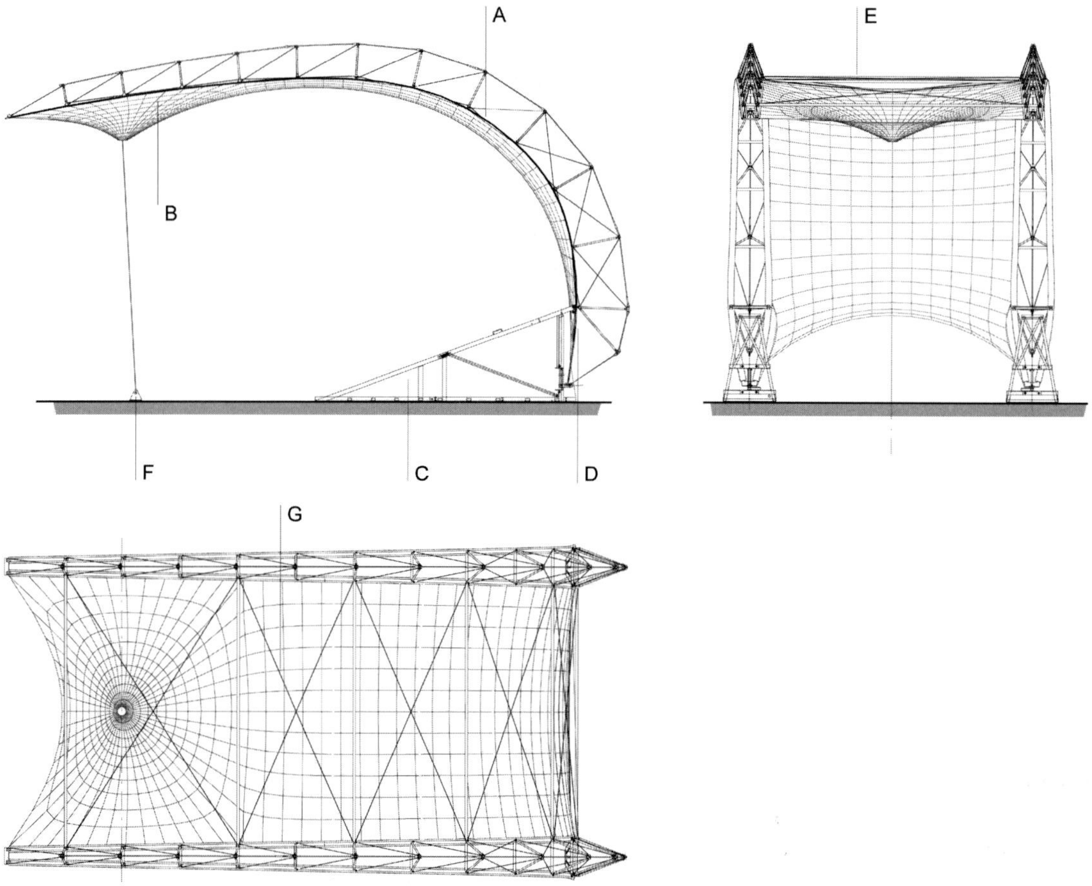

18.4 Full-size concept model

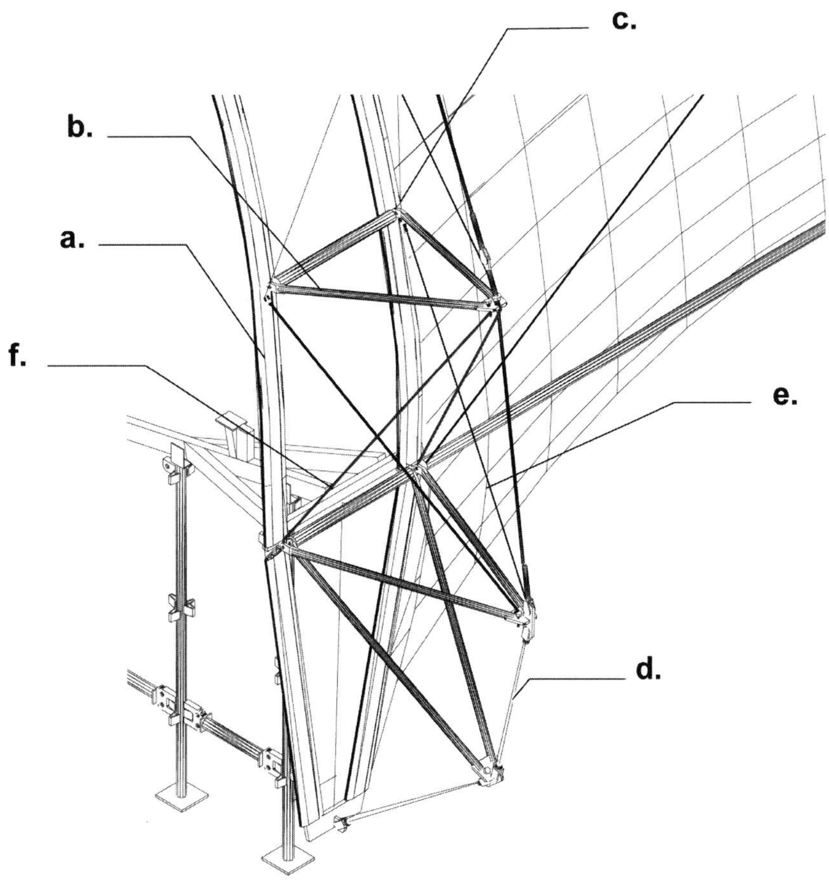

18.5 Truss components

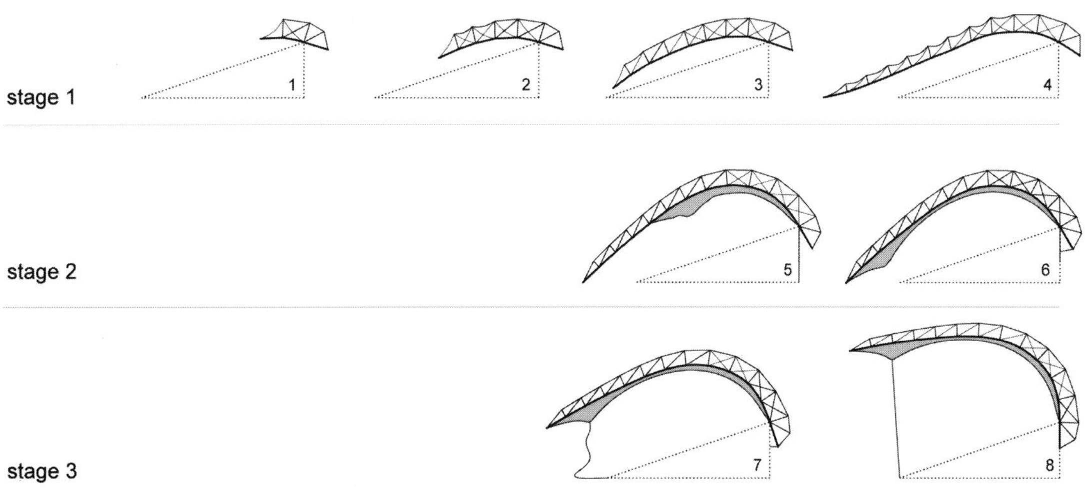

18.6 Erection sequence

Stage 2

Once each truss has been assembled, the trusses are lifted so that they deflect to their final working profile under self-weight. This occurs when the front end of the truss leaves the ground, thereby allowing the purlins to be inserted at a safe working height from the seating platform.

Stage 3

The membrane is inserted in luff grooves in the edges of the compression chord and locked off by pins at the front and back corners of the trusses. The tie-down cable and sway bracings are then fixed in position.

Stage 4

The truss is winched to its final height and the short cantilevered end is locked off to the seating support structure by a pin. The system is pre-tensioned by shortening the front tie-down cable to a predetermined length, stressing the compression chords of the truss against the curve defined by the A-frames and supporting cables.

General behaviour and loading

Truss – shape – improvements

During the erection the dead load provided by the trusses strains the compression chords reducing their radius of curvature. This has the effect of increasing the length of the top cable, which results in an increase in the tension in the cable. The fully pre-tensioned top cable combined with the A-frames and the diagonals limits the deflection of the compression chords and defines the working profile of the truss.

Pultruded glass-fibre sections were considered for making the compression chords in the prototype because of their lower material stiffness and hence the lower stresses caused by pre-bending the members into shape. However, previous experience had shown that the jointing of Glass Reinforced Polyester (GRP) components in such a complex structure would likely prove to be difficult and expensive. For this reason, it was decided to use aluminium instead. Because of the change in material and client requirements to increase the free span of the canopy from the 6-metre span of the concept model up to the 13-metre span of the prototype required changes to the original concept.

The flexural pre-bending of the aluminium compression chords into the fully deployed profile results in large initial material stresses – up to 100% of the yield stress, in the back-section compression chords. This was solved by manufacturing the compression members in this area of the truss with initial curvature close to the curvature expected in the fully deployed profile. Further pre-bending is then required to rigidize the structure in order to carry external loads. The solution is a compromise and further work is required to find a method that satisfies the structural, deployment and transportation requirements.

The simplest way to pre-stress the truss is to introduce a vertical load at the front of the cantilever. A tie-down cable was used which allows the structure to resist wind uplift. The final position of the tie-down in the concept model resulted from a combination of considerations between the truss and membrane. One consideration was the requirement to provide an anchorage at the base of the tie-down. As it would not always be possible to use ground anchors, ballast stored below the seating would be required. This needed to be as close to the front of the seating as possible, but this required enough space below the seating to locate it. This meant that the idealized position for the tie-down was some distance in from the front edge of the truss. Additionally, some mechanism was required to pre-tension the membrane in the lateral direction. Consequently, it was decided to tie the canopy through the membrane. Originally this was by cables attached to the front edge, but considerations of draining the canopy resulted in the membrane being formed into a downward pointing cone to which the tie-down was attached, thereby acting as both an anchor and drainage point (Figure 18.7).

The use of single forward bracing in the canopy ribs meant that under some loading conditions some of the bracing members became slack. This allowed the compression chord members to deform further due to the increased effective buckling length of the member. This problem has been eliminated with the use of cross bracing. Care, however, is needed in assembling the structure to ensure that none of the bracing cables are initially slack. Conversely, over-tight bracing causes other problems. Therefore it was necessary to find a reliable method of adjusting the bracing cables during manufacture.

Loads

In general a clear definition of the load cases is absolutely necessary in the field of lightweight and mobile structures in order to produce an optimized solution. The designer has to struggle with two basic problems. The mobility of the structure will mean that there are a large number of different load cases, which vary with the location of the structure. So, for example, the snow load can range between 0.75 and 2.00 kN/m^2. The next problem is that the majority of loads detailed in the codes do not represent the actual load cases that a particular structure will be subjected to. Of course it is possible to use reduced loads from the codes for temporary structures, but very often the regulations concerning the time of use or the dimensions and other boundary conditions do not accurately fit the problem. Therefore, it is essential to develop a load concept in cooperation with the client that closely defines a balance between the reasonable necessary and economical load cases. In the first stage of the development ARENA were looking for a structure that could be subjected to every load case (wind or snow) over the entire UK. It became very quickly apparent that this was not practical and that it would have a seriously detrimental effect on the rest of the design, particularly the deployability of the structure.

The pressure values for the canopy used in the calculations were obtained from the British Codes, which produced comparatively high-pressure coefficients in certain areas of the canopy. However, a parallel study using wind tunnels indicated that the actual values might well be much lower than the statistical values derived from the tables. Pressure coefficients were recorded on a 1:50 scale wind tunnel model. The model was set up to simulate a wide, multi-bay configuration because this generally represents the worst loading case. The size of tunnel did not allow a clear definition of the simulation of the boundary conditions of the model and it was thought that blockage in the tunnel might be distorting the results. Further evaluation is still continuing and options to reduce uplift at the leading edge of the canopy are to be investigated.

The full-size concept model has been observed in wind speeds of 12 metres per second. All of the critical components were monitored by strain gauges and no parts were unduly stressed.

Manufacturing considerations in the detailed design

From the outset, it was the intention to reduce the number of separate manufacturing operations to a minimum. It was also a consideration to try to develop the design so that it could be manufactured in-house by Arena using only simple tooling and a minimum of skilled labour. Where possible, the need for complex jigs to assemble components needed to be eliminated in order to reduce the initial set-up costs and operator time required during assembly.

The kit-of-parts approach that was adopted allowed the design to focus on using standard sections for

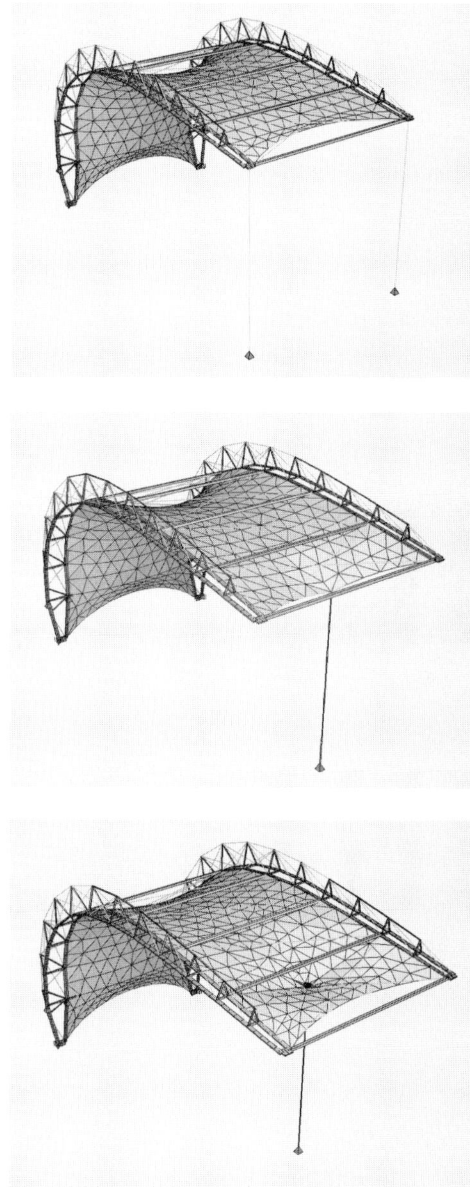

18.7 Variations of the position of the anchor point

the linear elements of the structure and a set of purpose-designed, integrated mouldings for the connections at the complex junctions between these. This would mean that the linear elements could be produced using continuous linear production methods and simply cut to the required length and angle governed by their position within the structure. The three-dimensional mouldings would then have to be designed in such a way as to be adaptable to the changes in angle between components in different parts of the truss and to the connection requirements at the alternative principal nodes. These elements could then be used as a self-jigging assembly in much the same way as a plastic 'Airfix' model kit is designed. This approach has the added benefit that most parts can be outsourced in sufficiently large quantities so as to reduce the individual parts cost. Cost estimates undertaken at the time indicated that the cost reduction could be as great as a third of the cost of traditional fabricated construction.

Material considerations in the detailed design

The approach to the detail design prefigured the materials that could be used. Aluminium became the material of choice due to the large number of alternative manufacturing processes available, its low specific weight and high strength. The wide range of high-grade alloys allowed the material properties to be selected according to the structural and physical requirements of the different components and the manufacturing constraints of the different moulding processes. The latter inherently contain high aesthetic value, which would be difficult to achieve in more conventional fabricated approaches and would be too expensive using machined parts.

Design and connection considerations of the principal components

There were three principal areas to resolve in the detailed design of the principal components, namely: primary compression chord, bottom node assembly and top node assembly. Although there were other components to detail, these areas contained the principal connections for all the major functional parts of the structure. The joints have been designed using a combination of conventional physical models and mock-ups, computer visualization and stereolithography models from computer-generated data.

Compression chord

The compression chord is a purpose-designed aluminium extrusion (Figure 18.8), manufactured by Hydro Aluminium. It contains the connection system for the membrane and jigging slots for the other mechanical parts. It was necessary to extrude the rib from high-grade aluminium to sustain the stresses induced in the material due to the pre-bending.

Top node

The top node (Figure 18.9) interfaces with the tubular uprights of the A-frame legs, the top cable and diagonal bracing cables. The top cable is in three sections to enable the truss to be split. Each section contains two or three locations where it interfaces with an A-frame. It was necessary to design a special connection for the intermediate junctions, as it was too expensive to split the cable at every interface. There were three individual configurations of the node within the truss, namely:

- junction at an intermediate connection of a top cable section
- junction with an end connection of a top cable section
- junction with bays in which there was a forward and backward diagonal cable

The node comprises a principal housing that is used to connect the tops of the A-frame legs. This housing then interfaces with a number of other cast components that are designed to resolve the

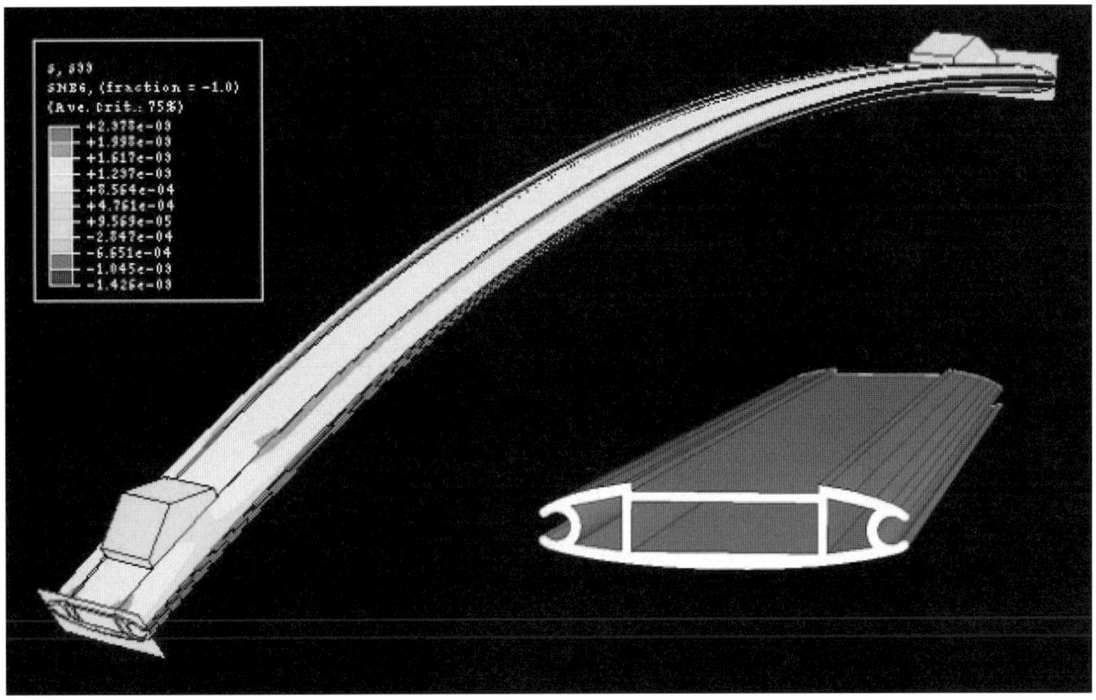

18.8 Compression chord

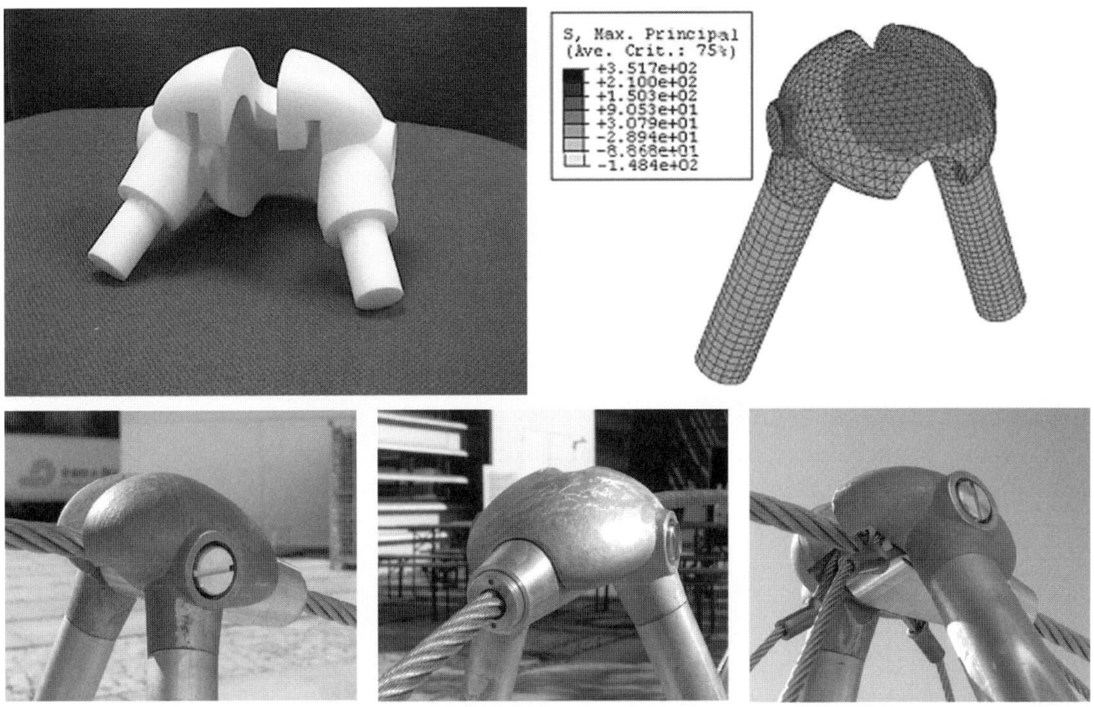

18.9 Prototype top node design

different connections between the components at each of the three alternative locations.

Bottom node

The bottom node (Figure 18.10) interfaces between the lower end of each A-frame leg, the horizontal member of the truss/purlin and the compression chord. It also picks up the forward and backward diagonals from adjacent bays. There were two principal configurations of this assembly, namely:

- junction at an intermediate connection of a rib section
- junction at a joint between individual sections of the truss

The node comprises a principal housing that is bolted directly to the rib. The housing encapsulates the horizontal member of the truss which is used as a pivot for a second casting which picks up the bottom of the A-frame legs, allowing the A-frame and top cable assembly to fold flat against the rib. In the second situation, a third casting interfaces in a socket in the back of the node. The geometry of the two castings forms a pivot, which allows the truss sections to be connected, and a pin is used to lock the joint. This allows the joint to transmit the tension and compression forces induced in the truss due to the pre-bending and applied loads.

Conclusions

The canopy is a highly efficient composite post-tensioned tensile structure with an anticlastic membrane supported by aluminium trusses (Figure 18.11). The design has focused on optimizing efficiency and safety during assembly and operation, increasing visual aesthetics, ease and simplicity of manufacture while minimizing overall costs.

A principal problem encountered in the development was the degree of innovation that was needed at almost every level of the design in order to achieve the high degree of integration necessary to optimize handling and manufacture. This meant that the design was left at an incomplete stage at the end of the project.

There are currently two principal areas where further work is required.

- the loading requirements of the structure, detail behaviour under different load conditions and the rigorous structural analysis and testing of individual components and assemblies.
- the further development of the node designs to increase integration and reduce weight.

For the future development it will be necessary to revise the original design based on more realistic loading conditions. It was apparent at the end of the project that the desire to have a temporary structure that would satisfy all design conditions, in all locations and at all times of the year, had had a serious detrimental effect on other areas of the design. The future development will need to be carried out as a partnership. One of the principal deficiencies of the original project was in the identification all the necessary parts of the supply chain and the appropriate partners for key areas of the project's manufacture. The structure currently remains in the prototype stage of development until sufficient funding and partners can be found to develop a production-optimized system.

References

Baglin, P., Wilkinson, K., Burford, N. and Smith, F. (2002) 'ARIES a Demountable Canopy', 5th International Conference on Space Structures, University of Surrey.

Burford, N. and Gengnagel, C. (2003) 'Ein Dach der besonderen Art' in *Mitteilungen der TUM 03/04*, Munich.

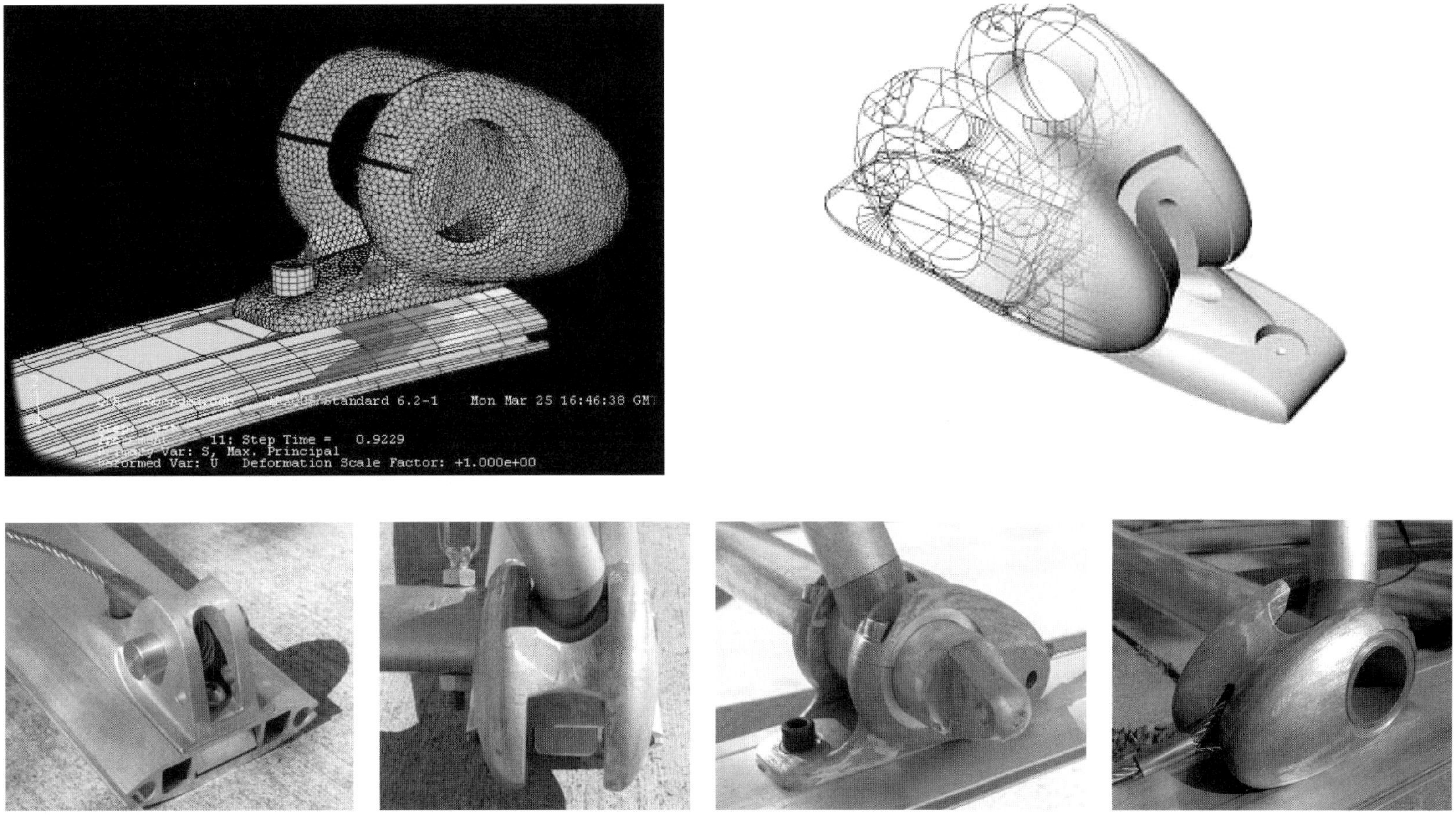

18.10 Prototype bottom node design

18.11 Erection at Munich in 2003

Biological Structures and Deployable Architectural Structures

Maziar Asefi
University of Liverpool

Introduction

The natural world has proved that it can be the best teacher in many disciplines. It can be argued that no new artefact can be created unless it employs principles that either exist in nature or have directly inspired the human mind. Biological structures are the most important natural features that have engaged architects' and engineers' minds. Various kinds can be found in nature, but how they may be translated into man-made architecture is unclear (or at least undefined) and requires substantial preliminary investigations and research. Buckminster Fuller believed that in nature a great technology could be seen to have been in existence for millions of years and he could not separate the concept of structures and technology from the natural world (De Varco, 1997). An expression of Fuller's understanding of nature can be seen in his design work on the geodesic dome in which the formal concept is the same as the structure of an insect's eye. Geodetic structures can also be found in many cell structures, for example the spherical groups of carbon atoms named Fullerene in his honour.

This chapter describes and explores some principles that exist in natural structures, which form models for the design of a particular sub-set of architectural structures that can be described as 'deployable'. Such structures can pass from a folded to an erect state and may alter their geometry during the process of deployment. In this process the structures' mobility is transformed into rigidity. These structures have the great advantage of speed and ease of erection and dismantling compared to conventional building forms. Buildings that incorporate these structures can benefit not only from the advantages gained by deployability but also from the unique quality and flexibility of the spaces provided.

Deployable architectural structures

In terms of kinematics, deployable structures are divided into two main groups – deformable and rigid-link. Examples of both these groups may be found in architecture. Deployable structures can also be used in both permanent and temporary buildings. The most obvious advantage that their application brings to architecture is that they can make buildings easier to erect, but they also have the potential to introduce flexibility and ecological efficiency, and to integrate technology and aesthetics into the production of dramatic structural forms.

An examination of the various categories of deployable structures shows that four principal factors play a main role in their design: structural mechanism, operation, transportation and materiality. It can be argued that when these factors are well considered the design of the resultant structure will meet a range of essential criteria including adaptability, multi-functionality and transportability. It is useful to treat these main factors as the basis for discussion of the possible beneficial effects of biologically based strategies on deployable architectural structures.

The design of intelligent deployable architecture that can respond to environmental changes is an important objective. The employment of smart technology could create structures that can adapt to humanity's changing needs and disparate functions and potentially make transportation and operation procedures easier.

Nature and flexibility in design: development and growth in natural biology

Examination of the process of biological growth and deployment reveals that awareness and response are the main features that are employed in reacting to internal and external changes. In order to understand the principles behind the process of growth and development in biology the basic organization of the cell can be examined. Each individual animal and plant starts life as a single cell and growth occurs by the process of cell division. The nucleus of each cell holds a length of DNA, which includes in full all of the coded information and instructions required for the construction and operation of the entire organism. When a single cell is divided into two cells, and this process continues exponentially, all the cells contain the same information including instructions for the design and growth of all parts of the organism. For instance, there are about 10^9 'letters' of coded information in every human cell, which are duplicated in every cell in the body (Calladine, 2000). During organism growth some cells may become the components of the eye, others of the heart and others of the liver, and use a very small amount of information saved in their length of DNA. The essential point here is that each cell finds its role in the organism because the cells send signals to themselves, which signify what they are to do and how.

Application of the procedure of deployment and growth in biological structures to the design and construction of structural material could lead to an attempt to make architectural and engineering structures more compliant and responsive. One of the issues that architects usually encounter throughout design is how they can minimize the negative impact of environmental and climatic changes. This issue is also an essential point in the design of architectural deployable structures, which are very often used as a temporary building solution in disparate locations and situations. Employment of the responsive biological structures model could lead to a new strategy in architecture and engineering in which the negative influences of the environment might be considered as positive influences. For example, the individual cell in a biological structure 'understands' its task through signals that come from other cells, which are associated with the coded information that exists in the length of DNA (which is the same in each cell's nucleus). These signals stimulate cells to perform their task. If such a process could be replicated in architectural structural components and materials as sets of coded instructions individual building elements might be able to perform pre-programmed functions. For example, environmental changes might act as a stimulus to actuate a specific coded instruction to carry out a planned task in response.

For example, environmental changes can be considered as signals that stimulate a central cell (sensor) that includes all the different coded information that exists in disparate bundles of cells in a material. On the basis of the signals that are received from the environment, the material can adapt itself to changing climatic situations. This might not only make the environment more satisfactory for users, but could also decrease the building's manufacturing cost as the need for duplication in structures and material for different situations and locations would be removed. This principle could also be applied to deployable mechanisms, the structures acting like live organisms that change their geometry and configurations as required.

As Calladine (2000) notes, there are difficulties in making a direct analogy between the process of growth and development in biology and structures due to the numerous points of differences between

19.1 The Golden Ratio in nature: (a) the Fibonacci Ratio and a sunflower, (b) cut-away section of a Nautilus shell

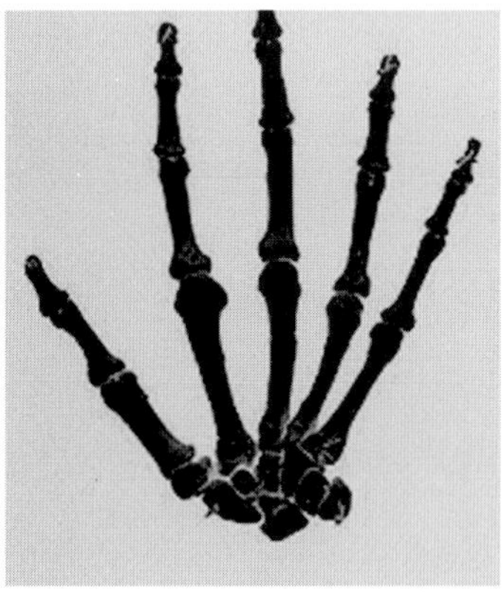

19.2 The ratios of the fingers of the human hand

these situations. However, regular use of reactive smart materials and components in building design is now becoming a reality and it is not unrealistic to suppose that further development will continue in this area.

Nature and material

Selecting a lightweight material, which is strong and yet flexible enough to form structure and cladding in diverse conditions, is a crucial factor in temporary building design. Cost, maintenance, flexibility, compactability and light weight are the main features to be considered when considering suitable materials for the covering of deployable structures. Biological structures include many materials that contain these features. For example 'Chitin' is a natural, thin, lightweight material, which covers the external body of insects. Due to its high strength and flexibility it can protect insects in different difficult situations. It is also damage-resistant and can be made in many colours and is consequently able to adapt to its environment (Yahya and Clarke, 2002). If such materials that exist in natural structures could be identified and duplicated for application in covering deployable structures a new form of adaptable architecture could result that would amount to a revolution in architectural design.

Natural geometry and deployable structures: Fibonacci Sequences (Golden Ratio)

The Golden Ratio is an irrational number, like other important numbers such as Pi. This means that it cannot be completely represented by our currently used number system, except as a formula $(\sqrt{5}-1)/2$. The Fibonacci Sequence is a series of numbers that are the sum of the previous two numbers (i.e. 1, 1, 2, 3, 5, 8 . . .). These numeric principles are not only seen in large numbers of living creatures (such as the daisy and the nautilus), but can also be found in a number of geometric shapes (Infinite Technologies, 2003) such as the hexagram, pentagram, equilateral triangle, golden rectangle and triangle.

If you examine the head of the sunflower there are usually twenty-one spirals in the clockwise direction and thirty-four spirals in the anti-clockwise direction; this ratio of 34/21 corresponds to the Golden Ratio (Figure 19.1(a)). A cut-away section of a nautilus shell shows that it grows in a spiral fashion. The distance of the lines drawn from the centre of the spiral that intersect the curve compare with the Fibonacci Sequence and the ratio of two numbers in this series equal the Golden Mean (Figure 19.1(b)). The Golden Ratio is also found in the structure of the human hand. The ratios of the length of successive bone segments in the fingers can be classified as a simple deployable structure that follows the Golden Mean principle (Figure 19.2).

Natural structures, particularly in plant species, employ the Golden Ratio as a packing rule in order to make the maximum use of available space. For example, the sunflower packs seeds according to Fibonacci's Golden Ratio – seeds grow radially from the centre of the bud and develop radially. An angular displacement for newly formed seeds is the Golden Ratio fraction of circle $(0.6180{*}360 = 222.5)$, which results in the best packing strategy (Esa, 2003). This packing method not only enables the daisy to produce as many seeds as possible but also decreases the amount of materials and energy that have to be used in doing so by creating the minimum size head. The wide variety of applications of the Golden Ratio in natural structures includes biological deployable structures, optimal packing, optimal proportion of spaces and forms. These provide ample models for their employment in deployable architectural structures.

Geodesic grids and natural structures

The geometry of some biological structures also shows the potential to be applied as lightweight architectural structures. As already stated, the concept behind the domes which Buckminster Fuller designed is the same as that found in the structure of an insect's eye (Figures 19.3(a), 19.3(b)), but this is not the only place where geodesic forms are found in nature. They also occur in many cell structures, such as the spherical groups of carbon atoms called Bckyballs as well as viruses, enzymes and even small organisms. A study of the geometry found in natural structures and the reasons behind their formation may provide engineers and architects with new directions for the design of structural forms and mechanisms.

A geodesic dome is a semi-spherical structure, which is comprised of a complex network of triangles. When the number of triangles increases, the dome more closely approximates the shape of a true sphere. The modern-day geodesic dome combines the formal and structural advantages of a sphere, which encloses the most space within the least surface, and is a stable structure resistant to internal and external pressure. In the late 1940s, Buckminster Fuller explored the concept of 'doing more with less' by creating a spherical building structure of unparalleled strength utilizing triangular surface planes (Buckminster Fuller Institute, 2004). The structure of an insect's eye is composed of many parts, all of which have to be held together and supported. This is achieved within a geodesic dome grid on the outside surface of the eye, which by its nature takes the shape of a hemisphere. The whole structural framework of the domed grid is held firmly in position and this in turn provides a stiff and strong support for the cornea and enables the insect's eye to function efficiently. In 1985, Richard Smalley and Tarold Kroto discovered this geometry in C60, a natural molecule structure, which is a form of carbon (Figure 19.3(c)) (Urner, 2004). This again proved Fuller's belief that in nature great technology has already been at work for millions of years. Experiments conducted between 1985 and 1990 established that this molecule is one of the most stable structures in existence due to its geodesic form and electronic bonding factors.

A geodesic dome is a suitable structure for transportable architecture because it can meet many of its principal design criteria. It covers the largest volume of interior space with the least amount of surface area, thus saving on material and cost. Its geometry has been used in mobile architecture both as deployable and non-deployable buildings. Engineers and architects such as Zeigler, Escrig and Calatrava have designed and constructed deployable structures based on the geodesic dome principles (Figures 19.4(a), 19.4(b)) (Escrig and Brebbia, 1996) and the structure still shows potential for further development as mobile architecture in the future.

Deployable mechanisms in natural geometry

Different mechanisms for folding and deployment can be found in biological structures. They can be categorized as planar, cylindrical, stiff and compliant (Vincent, 2001). Examination of the deployment mechanisms in natural structures confirms that nature generally applies folding properties for two main purposes, function and defence. An organism such as a plant may deploy or fold its components in order to protect itself from environmental attacks or to absorb the maximum amount of sunlight. Animals deploy their body structure for movement and to carry out tasks. The reasons for deployment in natural and man-made mechanisms are surprisingly similar and therefore the value of studying the existing systems is clear.

The example of the folding mechanism of Hornbeam leaves can be used to highlight the importance of the study of deployable behaviour in natural structures and their possible applications in architectural deployable structures. The leaves of

19.3 The geodesic dome and nature: (a) the structure of an insect's eye; (b) a geodesic dome, (c) a molecule of C60 (Fullerene)

19.4 Deployable X-frame icosahedron geodesic domes by Escrig, 2000

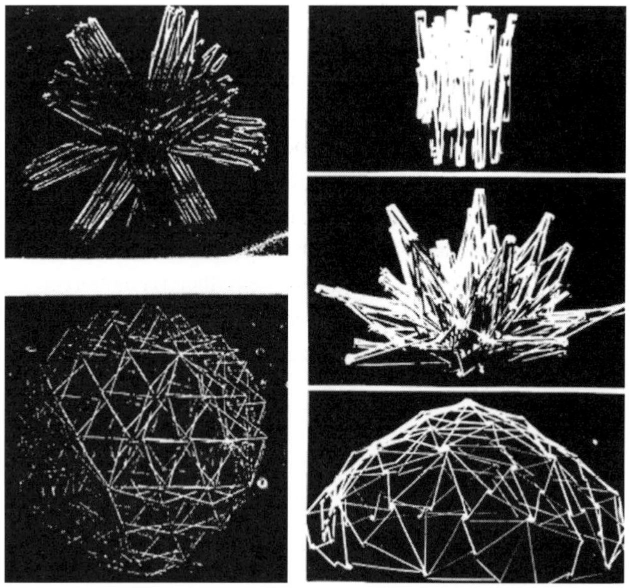

the Hornbeam have a straight central vein and symmetrically arranged parallel veins, which create a corrugated surface. When the main vein elongates and separates the bases of the lateral veins, the secondary vein rotates into the plane of the leaf, and in tandem the secondary vein rotates away from the main vein (Figure 19.5). Research on the angular arrangement of lateral veins shows that if the secondary vein angles range between 75 and 80 degrees because leaves can be folded more compactly than one with an angle between 30–45 degrees. This research has also confirmed that by using the small angle between the main vein and the lateral veins, then the largest deployed area will be achieved in the early stages of unfolding and that if the angle were to be increased, more energy would be required for the full unfolding (Kobayashi *et al.,* 1998). A study of the leaves of the Beech also indicates the same conclusion.

The results of such studies (and research on some flowers such as the poppy (Figure 19.6) whose main characteristics are the same as Miura-ori, a method for folding discovered by Koryo Miura from Tokyo University) may be used in architectural deployable structures to make them easier to transport. This mechanism has already been employed to improve the design of folded maps in order to make them stable in their fully deployed state to avoid a large number of movements and to prevent tears which start where two folds intersect. Poppy-type mechanisms have also been used in satellites, antennae and solar collectors (Bain, 1981). The combination of corrugated leaf patterns shows a potential to be employed in transformable tent and roof structures. The leaf-folding pattern can be arranged in different forms to produce polygons that could be built as folding membranes. By arranging them in such a way that they point either towards or away from the centre (Figure 19.7) these configurations can be used as folding roofs, solar panels and antennae (De Focatis and Guest, 2002).

The examination of folding mechanisms in natural structures, besides their possible applications as deployment principles in foldable architecture, may result in suitable information that can be applied to the design of intelligent materials for transformable structures. For example, the controlling factor in Sycamore and Maple leaf deployment is the use of different orientations in the cellulose micro fibrils in the cells' walls during the different stages of deployment (Vincent, 2001). This could be considered as a model for the creation of new smart materials that may enable whole structures to change form and strengthen foldable components during the folding and deploying process.

A striking imitation of a biological deployment mechanism derived from the animal world can be found in Renzo Piano's IBM Ladybird Travelling Pavilion. Though this building resembles an insect, its structural mechanism was inspired by the unfolding skeleton of a bat's wing (Figures 19.8(a), 19.8(b)) (Buchanan, 1997). Although this project was not constructed, it indicates the potential of biological structures to be employed in deployable temporary building.

In Calatrava's dynamic designs, inspiration from biological structures for both the entire geometry and the deployment mechanism can be found. The most impressive feature in his work is that he integrates biological structure, technology and aesthetics, producing structural forms to create buildings that are both visually striking and structurally daring. The planetarium in Valencia can be perceived as an abstract sculpture, but it also looks like an eye with metal sides that deploy and retract like eyelids (Figure 19.9). Milwaukee Art Museum, a recently completed Calatrava project, is enclosed by a large roof structure, which opens and closes like the wings of a bird to allow people inside the museum to observe the changing light of the site in the course of a day (Figure 19.10). The inspiration of biological structures is used in these buildings to

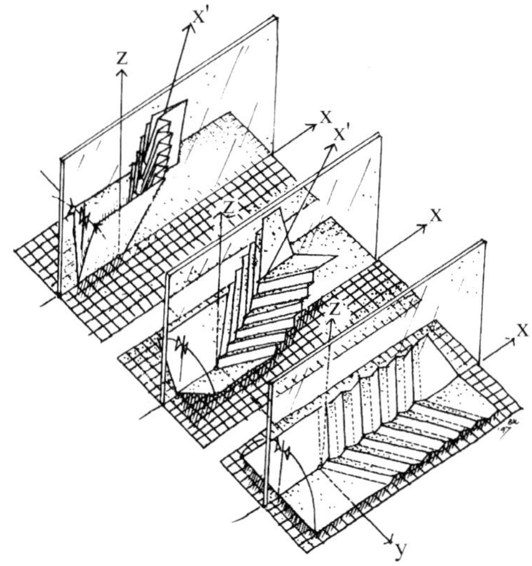

19.5 The geometry of a simple expanding Hornbeam leaf (Kobayashi *et al.*, 1998)

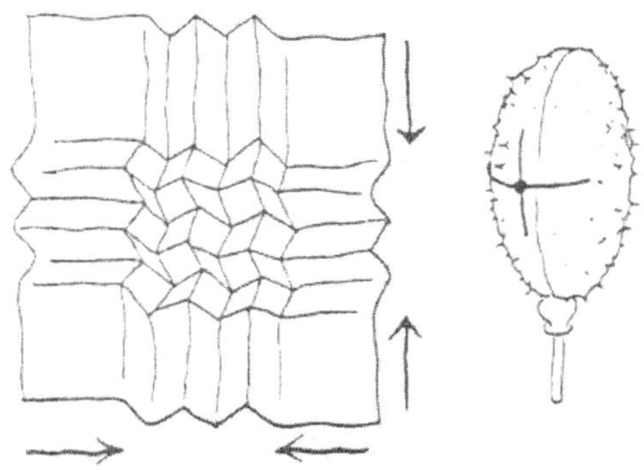

19.6 The poppy petal compresses itself as it grows in the bud (Kobayashi *et al.*, 1998)

19.7 Proposals for foldable membrane based on the assembly of the model of foldable tree leaves

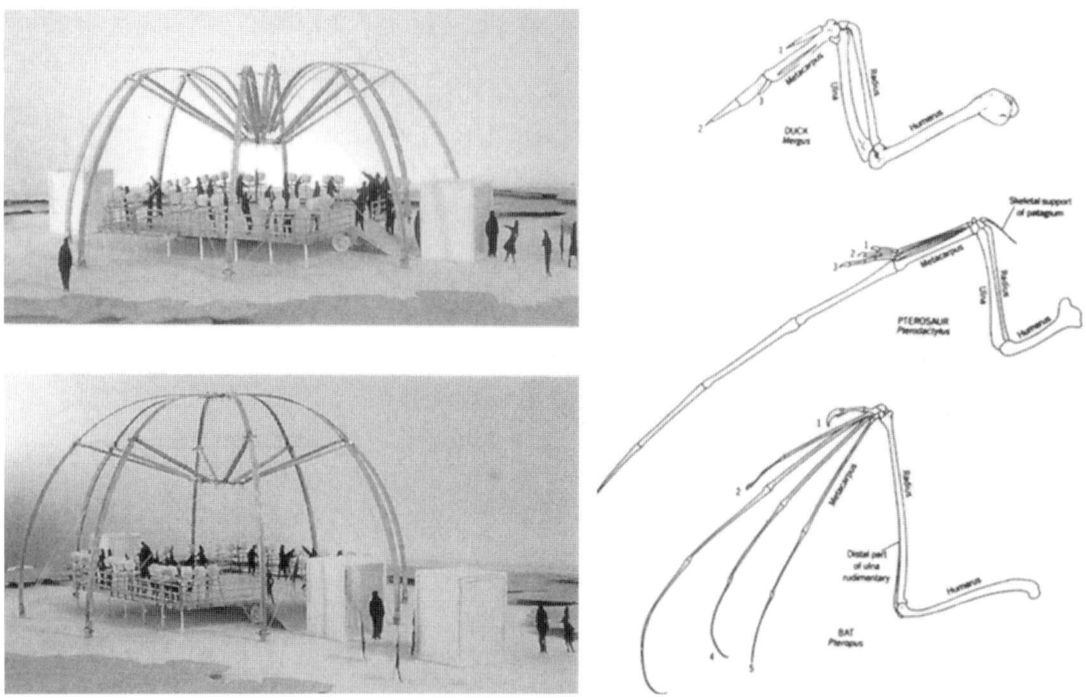

19.8 Ladybird Travelling Pavilion by Renzo Piano: (left) model view showing the structure of the pavilion unfolding (Buchanan, 1997); (right) bat wing skeletons that inspired the structure of the folding arches (Buchanan, 1997)

19.9 Valencia Planetarium by Santiago Calatrava

19.10 Milwaukee Art Museum by Santiago Calatrava

add flexibility and responsiveness. Here, deployability accentuates the sense of space at different times and in different locations.

Conclusion

In the design of deployable architectural structures, architects and engineers encounter three main issues. One of these pertains to general deployability and includes easy erection, transportability, sustainability and the need to create a lightweight, compact design. The second is associated with general architectural issues such as form, function and respect for human scale. The last and perhaps the most important issue is related to the character of deployable structures as temporary buildings and includes multi-functionality, response to environmental changes, adaptability to disparate situations and locations. Consideration of all these issues shows that the design of deployable architecture is much more complex than conventional building types. The contemplation of nature in certain biological structures reveals that there is a potential to benefit deployable architectural structures in all three of these design criteria. Seasoned architects and engineers may extract and translate the ideas and principles behind the process of growth and development in biological structures to make foldable architectural materials and component parts more adaptable to environmental changes. Moreover, exploring significant principles in living nature such as the Fibonacci Ratio can be applied to various functions such as deployability, optimal packing, appropriate proportion of space and geometric forms, and modular design, may help designers to solve these complex, yet relevant design issues. Some foldable buildings may be able to adapt the entire geometry of a biological structure to make architecture more lightweight and geometrically efficient (for example, the geometry of an insect's eye is in the form of a geodesic dome). A study of folding mechanisms in biological structures (for example, the leaves of the Hornbeam and Beech trees) shows that the process of deployment and packing are translatable not only into complete building structures but also as architectural component parts. The successful imitation of biological structures depends on how and to what extent they may stimulate our imagination and creativity. Further research in this field would benefit from a closer sharing of the expertise of biologists with architects and engineers and this collaboration could be exploited to increase effective solutions for the design of deployable architectural structures.

References

Bain, I. (1981) 'The Miura-Ori Mao'. http://www.britishorigami.org.uk/theory/miura.htm (accessed 10 July 2004).

Buchanan, P. (1997) *Renzo Piano Building Workshop: Complete Works.* London, Phaidon.

Buckminster Fuller Institute (2004) 'Introduction to Geodesic Domes'. http://www.bfi.org/domes/ (accessed 5 August 2004).

Calladine, C. R. (2000) 'Deployable Structures: What Can We Learn from Biological Structures?' in S. Pellegrino and S. D. Guest (eds), *IUTAM-IASS Symposium on Deployable Structures: Theory and Applications*. Dordrecht, Kluwer Academic Publishers, pp. 63–76.

De Varco, B. G. (1997) Invisible Architecture 'The NanoWorld of Buckminster Fuller'. http://members.cruzio.com/~devarco/invisible.htm (accessed 7 August 2004).

De Focatis, D. S. A. and Guest, S. D. (2002) 'Deployable Membrane Designed from Folding Tree Leaves', *The Royal Society*. http://www.deployables.net/ddf/academic/files/papers/depmem.pdf (accessed 23 November 2004).

Esa (2003) 'Deployable Packing Structures.' http://www.esa.int/gsp/ACT/ACT_Web/Subjects/Bio/struct_mat/struct/structure_S.htm (accessed 23 November 2004).

Escrig, F. and Brebbia, C. A. (eds) (1996) 'General Survey of Deployability in Architecture' in *Mobile and Rapidly Assembled Structures.* Southampton, WIT Press, pp. 3–22.

Hanaor, A. and Levy, R. (2001) 'Evaluation of Deployable Structures for Space Enclosures' in *International Journal of Space Structures*, 16(4), pp. 211–230.

Infinite Technologies (2003) 'Geometry in Natural World'. http://www.infinitetechnologies.co.za/articles/geometry1.html (accessed 12 August 2004).

Kobayashi, H., Kresling, B. and Vincent, J. F. V. (1998) 'The Geometry of Unfolding Tree Leaves' in *Proceedings – Royal Society of London,* pp. 147–154.

Urner, K. (2004) Geodesic Domes, 'The Geometry'. http://www.grunch.net/synergetics/domes/domegeo.html (accessed 5 August 2004).

Vincent, J. F. V. (2001) 'Deployable Structures in Nature' in *Courses and Lectures – International Centre for Mechanical Sciences* (412), pp. 37–48.

Yahya, H. and Clarke, A. (2002) *The Design in Nature*. London: Ta- Ha. http://www.harunyahya.com/designinnature0.php (accessed 10 October 2003).

Projects

LOT-EK, Ada Tolla and Giuseppe Lignano (New York, USA)

Robert Kronenburg
University of Liverpool

20.1 American Diner #1: transportable restaurant made from two shipping containers (1996)

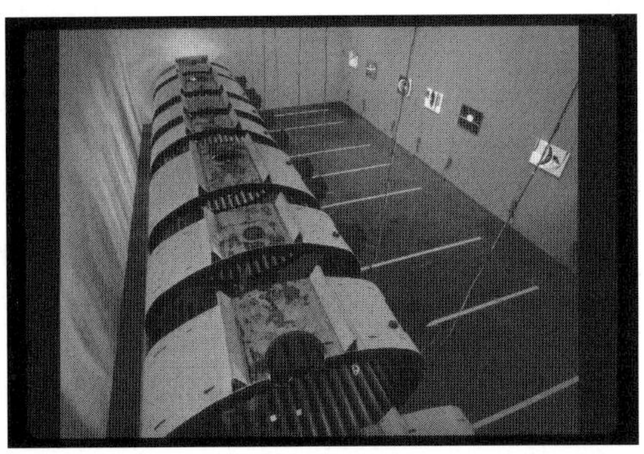

20.2 TV-Tank: a mobile television lounging tube built from a petroleum trailer tank most recently relocated at John F. Kennedy airport (1998)

In their curricula vitae Giuseppe Lignano and Ada Toller humbly describe their presentations about their work as lectures – they are not. To attend a LOT-EK presentation is an event. You sit in the darkened auditorium (a little darker than usual – this should be a warning) and wait for the casual joke, the personal anecdote, at least a brief introduction. It doesn't come. Instead a series of images begin to pass across the screen – each one intense and interesting yet there for only an instant. A voice is heard – Giuseppe's or Ada's – not sentences but words, not descriptions of what you are seeing but still . . . connected to it. The images are nearly exclusively of urban environments focusing on the things that we almost do not see, vehicles, hoardings, service gantries, road overpasses, bill boards, lighting poles, cable and pipe runs, television screens, advertisements again. Then you see a theme in the images and the words (which it becomes clear are in alphabetical order) become meaningful – *aberrant*, *abnormal*, *abstruse*, *absurd*, *alien*, *anomaly* – and begin to add information to what you are seeing. Suddenly there is a pause and the images are of a LOT-EK project and the words become sentences relating an absolutely straightforward description of what it is, what it does and how it was made. No theoretical wondering (or wandering). Then we are back into the images and words until the next project interjects.

This communication of LOT-EK's ideas and work is without doubt a performance rather than a lecture, and some may consider this a reflection of their apparently uncertain image in the critical design world as artists . . . or as architects. The presentation certainly has a considerable amount of style, and impact, and focus, but it is also an extremely direct way of communicating an essential characteristic of LOT-EK's work – its basis being in the careful observation of what is already around us and appropriating that for its aesthetic, functional and creative properties. In their presentation, Giuseppe and Ada let you see the built world through their designer's eyes and they do it better than anyone else I have come across. Through the project they show you the results of their observations and their subsequent efforts, without comment, without excuse. Their work, set in this context of everything that is already out there, can be judged by its physicality, its operation, its presence and of course its image.

LOT-EK's fresh eye might result from their presence as foreigners in their adopted land – they were both born and raised in Naples and graduated from the Università' di Napoli before completing postgraduate studies at Columbia University in New York. If so, they are not the first immigrants to reinvigorate their new home with an enlightening perspective. They describe the built environment of North America as a new natural landscape – clearly human-made, but one in which the familiar pattern found in the biology of genus and group can be attributed to the objects that surround us. The fact that these objects have value and beauty beyond their pragmatic usefulness is their theme. Though LOT-EK's appropriation and reuse of these objects has so far been primarily for art galleries and knowledgeable clients it does not appear elitist, perhaps because it still maintains solid links with the objects'

© Paul Warchol

20.3 Morton Loft: a reinvented dwelling utilising a petroleum trailer tank to provide private areas in a fourth-floor apartment (1999)

20.4 Mixer: a media cocoon built for the Henry Urbach Architecture Gallery, New York in 2000

20.5 Student Pavilion: University of Washington multi-functional space made from the fuselage of a Boeing airliner (2000)

20.6 MDU (mobile dwelling unit): prototype relocatable dwellings made from standard shipping containers (1999–2004)

20.7 CHK (Container Home Kit): mobile prefabricated house design utilising standard shipping containers (2004)

original use and so their objectives – to make us think again about what is beautiful and useful and what is not – are clear.

LOT-EK's work is apparently tough, resilient and long lasting. The fact that it is made from recycled materials and objects is significant but not in the conventional ecological way. The designers are happy that their work has this very real connotation with making something new out of something old, but it is the architectural potential of the objects they choose that concerns them. The base products from which their designs emerge are chosen very carefully and the juxtaposition with their new use and new surroundings is important in generating the aesthetic. In most cases this relies deeply on the idea that the new architecture is transportable because the very act of finding a recognisable structure in an unfamiliar place makes one think of how it got there and where it came from. LOT-EK's work takes on the industrial world and makes something new from it – a beautiful and surprising place that is surprisingly user-friendly. Many of their designs have been about creating an intimate human-sized space, for dwelling (Guzman Penthouse, Morton Loft, MDU), for relaxation (Mixer, TV-Tank) or for social gatherings (American Diner #1, University of Washington Students' Pavilion). Yes these designs are tough, aggressive even, but their very toughness leads to understanding that they exist for use and interaction. They are also completely non-precious, interesting and ingenious and so connect with two linked experiences that are all too rare in architecture – surprise and joy.

Office of Mobile Design, Jennifer Siegal (Los Angeles, USA)

Robert Kronenburg
University of Liverpool

Jennifer Siegal is founder and principal of the Office of Mobile Design (OMD). Founded in 1998, the Office of Mobile Design concentrates on developing 'mobile' architecture – designing and constructing portable, demountable and relocatable structures. OMD's focus is on finding unique, dynamic, environmentally sustainable solutions to both unconventional and typical design problems. Its practice involves providing design services to public agencies, non-profit organizations, schools, museums, commercial businesses and private interests. Recent projects vary dramatically and include the PuppetMobile, a travelling theatre and teaching facility for Cal Arts' Cotner Center for Puppetry and the Arts, as well as the Mobile ECO LAB, used throughout Los Angeles County by the Hollywood Beautification Team to teach school-aged children about environmental issues.

OMD's work deals with real problems in pragmatic yet inventive ways. Much of the design work focuses on user-led problem solving – be it low-cost housing, mobile education facilities or a travelling film kiosk and ice cream parlour. Though Siegal embraces modern materials and techniques and is aware of the impact of new technology on both the way we build and the way we live, her approach has been primarily aimed at creating designerly solutions that are affordable. The problem of providing architectural resolutions that deal with the added issue of mobility is particularly demanding. Adding this significant performance criterion can make the capital costs of workable solutions expensive, regardless of the fact that their operation achieves so much more for the money. Siegal's projects indicate an unprejudiced approach to the use of technology and materials: if new technology is essential to solve the problem then she embraces it; if not, reusable and recyclable materials and components are fine too.

Her Swellhouse project combines smart building technologies with cost-efficient prefabrication methods to create a deployable dwelling that can deliver the highest standards of living anywhere. Her Seatrain house, built in Los Angeles, uses recycled storage and grain containers to deliver the same high standard of living but in a completely dedicated solution using unconventional resources and building techniques. This awareness of real-world design issues, extending beyond the realm of the large corporate budget, makes her mobile building solutions of particular interest. OMD's production line projects are an aesthetically aware continuation of the mass-manufactured industry, which are created to provide more for the budget as a result of a specifically design-aware approach. Siegal's one-off projects are dedicated solutions that reinvent the way we might use familiar facilities if they were infinitely mobile yet still fundamentally practical.

Wheels are an important part of OMD's design approach, which appears to be that any city environment can be made more usable and more dynamic if it can be hitched up, towed, pulled or driven from place to place. The standard, readily available vehicle provides a blank canvas for attaching a multitude of functions from a bicycle that becomes a shop to a 12.2-metre (40 foot) trailer that becomes a school, surgery or IT facility. This utilization of the existing to

21.1 Mobile ECO LAB built for the Hollywood Beautification Team as a transportable classroom focusing on ecological issues

21.2 Swellhouse; mass-produced building techniques utilising environmentally sensitive materials to create a high-quality deployable residence

21.3 Seatrain residence, Los Angeles: overhead view of the site – the recycling yard to the right provided both inspiration and resources for the project

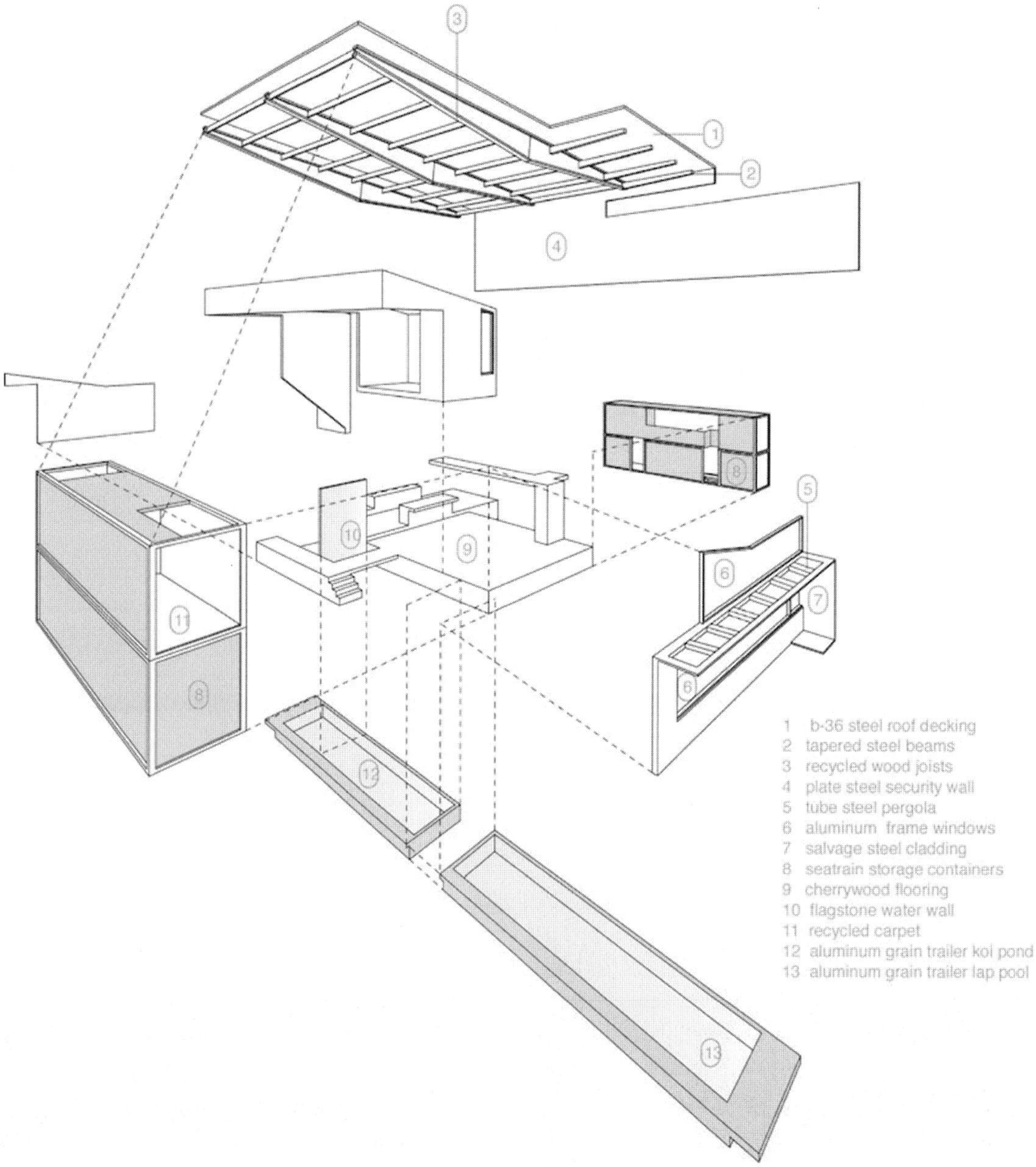

21.4 Seatrain: drawing showing the construction strategy

21.5 Seatrain: view from the garden towards the main house

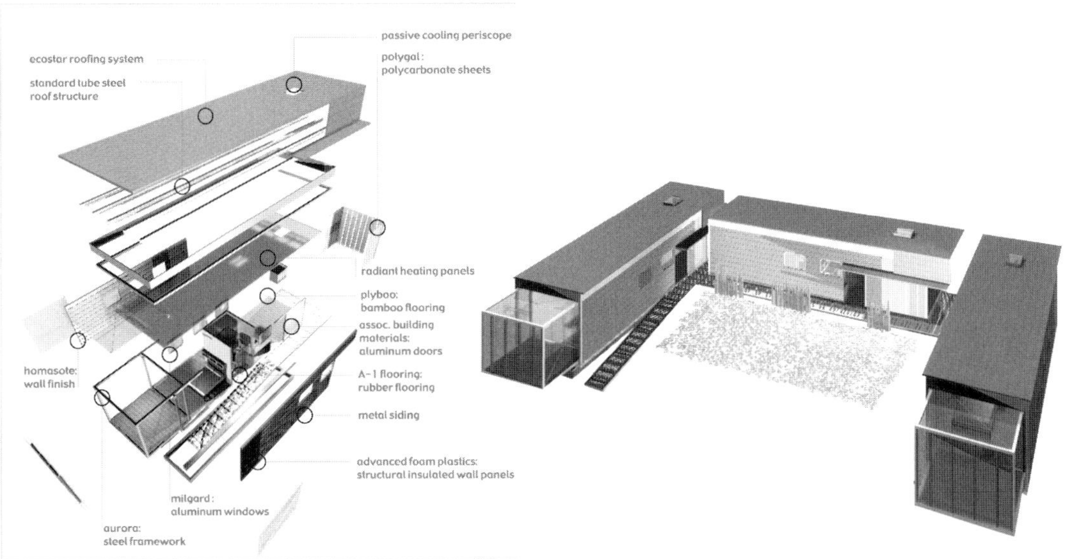

21.6 Portable house: an adaptable affordable dwelling that can be erected in different combinations and layouts to provide variety and quality of experience in a low-cost housing environment

21.7 Proposal for a sustainable community in Los Angeles

create the new is a key element in Siegal's work. In 2002, OMD was commissioned to design a 2.5-acre sustainable development in downtown Los Angeles comprising more than forty housing units. Instead of levelling the site and creating a new environment from scratch, Siegal chose to create an interactive system of factory-produced units that could be deployed among flexible communal gardens and workspaces and retain and renovate the existing warehouse structures into live-work dwellings. The new units are a redesign of the typical mobile home, using conventional manufactured housing construction techniques to make contemporary and desirable, yet low-cost houses. OMD's design keeps to the essential standard sizes necessary for economic production, but varies the components (including unusual but ecologically friendly materials such as bamboo) and the organization and disposition of units on the site (including the provision of two-storey buildings) to create a unique city neighbourhood.

As well as a practitioner, Siegal is a professor of architecture at Woodbury University in Los Angeles where she has attained national recognition and received numerous awards for her community-based design-build architecture studios. She was awarded the Association of Collegiate Schools of Architecture Faculty Design Award in 1999, 2000 and 2004, and the Westside Prize from the Los Angeles Westside Urban Forum in 1999, amongst other awards and distinctions. Her work has been published in *Architectural Record*, *Wallpaper*, *Domus*, *Dwell*, *Time, Esquire, Elle, Casamica, Metropolitan Home*, the *Journal of Architectural Education* and the *New York Times*. Her work has been exhibited at the Cooper Hewitt Museum in New York and the Walker Art Center in Minneapolis. Her book *Mobile: The Art of Portable Architecture* was published in 2002. She was a Loeb Fellow at Harvard in 2003. She is the series editor of the upcoming publication *Materials Monthly*.

OMD's work explores urban community-based theories and practices and links it to the practical boundaries of contemporary architectural design. This work is not so much concerned with mobility as with the design demands that make mobility necessary. Siegal states she is interested in studying the permutations of new materials, design styles and transportation methods that allow for the creative applications of mobile architecture. However, it is certain that she is just as interested in the people and the places she creates by doing this.

Biomimesis in Architecture: Inspiration for the Next Generation of Portable Buildings

Semra Arslan and Arzu Gonenç Sorguc
Middle East Technical University

Human beings, in accordance with nature, have a tendency to discover and learn from their environment. Thus, human-made technical developments sometimes directly result from and synthesize these observations of nature. In our observation/learning/designing process, we have experienced adaptation and developed skills to provide our needs by imitating, interpreting and using the opportunities of nature. Consequently, similarities observed between human-made structures and the structures of nature are unavoidable.

Biomimesis (from *bios*, meaning life, and *mimesis*, meaning to imitate) is a new science that studies the best ideas that can be drawn from nature and then imitates their design and processes to solve human problems. This science seeks new technologies useful for farming, feeding, manufacturing, healing, designing, constructing and even aiding in the operation of the world's corporate businesses in the information age. The concepts of biomimesis were first employed by Leonardo Da Vinci in the fifteenth century in his wide-ranging studies, from engineering to medicine.

In today's world the answers to difficult questions (such as how we harness energy efficiently, how we achieve lightness in structure, how we make economic use of materials, and many more) are being sought in biological research laboratories by interpreting different aspects of biomimesis. It is a multidisciplinary subject involving many fields ranging from electronics to engineering, from informatics to architecture.

Scale, function and process may be different in nature; however, design constraints and objectives are very similar. For example, functionality, optimization and cost effectiveness are all factors that impact on both the natural and the human-made world. Therefore, it is not surprising that humanity has always admired biological structures and often been inspired by them, not only their aesthetic attributes but also their engineering and design quality and efficiency. Significant engineering structures such as the Eiffel Tower and the Crystal Palace are designs that borrowed their ideas and principles from nature. Design principles developed by Buckminster Fuller and Frei Otto for application in new structural and construction systems exhibited a deep concern for biological phenomena. More recently, Santiago Calatrava has explored structures found in nature that have the capability of motion.

Life itself is integrated with the continuous motion found in all organisms from the single cell to the most complex system, from subatomic particles to galaxies. When motion ceases, life ceases. All living creatures have the ability to move – an integral factor of life. For instance, animals need to move to hunt or migrate – plants move according to climatic conditions. All the species have particular mechanisms tuned to their own need for motion. As living beings, humans also possess this need to move to continue their lives. Since ancient times, in order to create a place for living, for accommodating vital activities, humans developed architecture. Throughout history, humans also relocated as necessary, carrying their objects and possessions with them.

These objects include portable buildings, which have been in use since humankind first began to build. Traditional building forms such as the tent, tipi and yurt have sophisticated constructional forms, which reveal similarities with structures found in nature like birds' nests, ants' towers and spider webs.

In today's world, despite most buildings being 'static', the design examples of Otto, Piano, Ando and many others still aim to provide the type of buildings that demand portability and take nature as a source of inspiration. It is now necessary to ask how we can be inspired by nature to design new portable structures and/or buildings and how biomimesis can contribute to those new proposals.

Dockable Dwelling

Matias Creimer
OTIS College of Art and Design

Visiting a construction site reveals both the beauty of hands-on craftsmanship and its inefficiency. Given the low-tech nature of many trades involved in house construction, the latest technological developments that reshape many other industries are of little or no help. While developments in assembly-line robotics and qualified labor bring down the price of items that involve sophisticated designs, such as phones, computers, and cars, the construction of houses is still at the mercy of tradesmen working under stringent site and weather conditions. The critical size restrictions for the shipment of high-tech factory-assembled buildings, or building modules, are determined by trucking regulations and the transportation infrastructure. Prefabricated houses fall within these critical dimensions which indicate that the obstacle preventing affordable factory-built houses from being more wide-spread is not of a technical nature.

The *dockable dwelling* project, in an effort to bring design and construction efficiency within the affordable housing market, proposes a factory-assembled modular system that conceives the components of a house as similar to the middle cars of a train. Borrowing from NASA's approach to "space station modular docking," where minimizing space-walking time is critical, the proposed collection of modules can be fully built in a factory with almost no on-site assembly requirements except the portions that allow leveling and plugging of services.

In order to promote diversity and competition, this project proposes the adoption of a single "universal docking port" system that will ensure compatibility among different modules that can be manufactured and designed by different companies. All modules arrive fully insulated, finished, and equipped with ducts, wires, and pipes converging to specific points on the faces of the universal docking ports for service hook-up. A single photovoltaic-insulating membrane is proposed for energy efficiency, waterproofing, and thermal protection. At the end of the buildings' life cycle, the modules are shipped back to recycling facilities, similar to those in the automotive recycling industry, to be disassembled and recycled in a 'cradle-to-cradle' fashion.

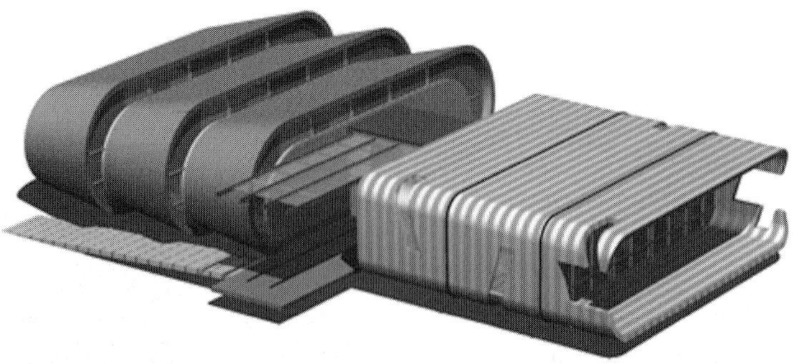

Crysalis: A Portable Personal Shelter

Jessica Davies, Yvonne Cheng, Johanna Doukler, and Tamsin Ford
Ryerson University

This second-year student design project was conceived, implemented, and constructed within approximately one month in 2003 as part of a university curriculum directed by Associate Professor Filiz Klassen. The program for *crysalis* was determined by the design team after taking into consideration the given project constraints such as minimal material choice and ease of deployability.

The project title makes reference to the word "chrysalis," which is the name given to the pupa of a butterfly or moth when enclosed in a cocoon. Taking inspiration from animal architecture, the team used plastic bags – a readily available material – to create a portable personal shelter which was intended to induce relaxation and calm. Its form was a suspended cocoon-like structure that referenced the stillness of our first natural shelter – the womb. Through *crysalis,* the team sought to explore ways that the individual might relax away from the high levels of chronic stress and anxiety most commonly experienced today in urban societies. The constant exposure to and inability to escape from these pressures prompted the team to consider a portable relaxation module to accommodate the increasingly nomadic and transitory nature of many living and employment conditions.

Crysalis is designed to function as a semi-permanent nest where the user can lounge, relax, catnap, think, or meditate in live/work environments such as an office, studio, home, or retreat. A front hatch allows one to rest with a view, or alternatively to be completely enclosed. *Crysalis* can be suspended indoors or outdoors by being attached to a beam, tree, or column. It collapses into itself for ease of transportation. The project reuses (and so recycles) a readily available and abundant material, the plastic bag. Approximately 3,000 low-density polyethylene plastic grocery and garbage bags were collected and salvaged for the project. Unique, material-specific joinery methods were developed to enable the team to create the structure within a two-week time period. When a light source is located inside, diffused light filtrates through the weave of the bags, softly illuminating the interior space.

In responding to the material constraints required by this project, the team was forced to investigate unconventional materials. The choice of materials, therefore, became strongly intertwined with the potential character and shape the shelter took on. The choice of plastic bags satisfied this constraint in many different ways, offering an extremely strong, weaveable, collapsible, cleanable, and accessible material. In finding a second value for a material that would normally become waste after one usage, the team was able to extend the lifespan and usability of plastic bags. As well, this project challenges the notion that this sort of plastic is a weak, cheap, unappealing material. *Crysalis*' notable woven texture, brightly colored decoration, and soft feel present a completely different social and physical interaction with the material.

In·ter·tex·ture: Weaving Kinetic Structures

Xenia F. Diente and Elise Knudson

We live in a world populated by structures – a complex mixture of geological, biological, social, and linguistic constructions . . . we cannot help but interact in a variety of ways with the other historical constructions that surround us and in these interactions we generate novel combinations, some of which possess emergent properties.
(De Landa, 1997)

This presentation In·ter·tex·ture at the Transportable Environments 2004 conference in Toronto, consisted of a three-part performance demonstrating the physical manipulation of geometric forms through space and time. Diente's kinetic sculptures became *Ptolemaic puppets* as Knudson's hands and body moved. She kneaded the contraptions from hyperboloid to tauroid, or pulled against a tensile fabric, and she challenged the audience to follow in the act of discovery. The quotation from Manuel De Landa above informed their collaboration in which they weaved together lightweight structures and movement. Geometry and movement were integral to Diente and Knudson's project and they took references from the exercises of master alchemists like structural engineer Frei Otto, inventor Chuck Hoberman and modern dance choreographer Alwin Nikolais.

The collaborators first met at the Atlantic Center for the Arts *WindArt* 2003 residency program. Individuals who joined the residency were well versed in kite design, aerial photography, aerial performance, and other disciplines. Diente's work illustrated the morphological properties in simple geometric forms through prop design and Knudson's experiments in stripping the human body from its normal context by using props and materials brought them together. At the residency they began weaving their disciplines and media, and thus collapsing the boundaries between geometry, sculpture, and dance. They employed simple technology and motion to enhance interaction, rigor, and potential energy inherent in the material makeup of Diente's sculptures and Knudson's work on the human body. They found that their collaboration gave way to geometric patterns that often took on metaphoric symbolism.

At the Transportable Environments 2004 conference, their presentation took place in the atrium of the host academic building surrounded by walkways on the ground and second floor, temporarily serving as improvised lookouts and seats. Three 5–7-minute long sets were cued by music from a small boom box while the site-specific performance unfolded from small travel bags and cardboard boxes hidden in the corner behind the door. The first performance broke the interaction between the audience and performer. Knudson's dance on the ground floor started with a costume in a tensile fabric that was attached to the second-floor railing among the audience. When the dancer moved, so did the two-story fabric costume. For the second performance, Knudson manipulated one of Diente's kinetic sculptures, by transforming what looked like a bundle of sticks to a bouncing tauroid in the course of her dance, and created *yurt*-like forms. The third performance engaged the audience directly with a game of 'Simon says' using six of the same sculptures from the second performance. The changing repetition of the form by six volunteers created a kaleidoscopic impression when viewed from the balcony above. The ground-floor performances, open to public passage throughout, utilized the interruption of patrons entering and exiting the building as unplanned but welcome components of the event.

An unfurled piece of paper contains the memory of the origami bird that was once manifested in its geometric creases. In a similar way, a molecule of DNA contains instructions for building the whole animal. Although working in different media, Diente and Knudson are both guided by a philosophy that converges on the behavior of sculpture/structure as a whole as dictated by the properties of its most basic building block. When Diente builds a kinetic sculpture, she begins with an equation, or a ratio of parts, which then dictate the shape and motion of the finished product. When Knudson choreographs a dance, she begins with a set of parameters that inform all of her movements that follow. Since "the rule" (the equation or the parameters put on the movement) is the starting point, neither necessarily know what the seed will grow into at the outset of a project. This is why their work is an act of chance and discovery.

References

Atlantic Center for the Arts *WindArt* 2003 (http://www.subvision.net/aca-windart/).
De Landa, M. (1997) *A Thousand Years of Linear History*. New York: Zone Books.

i-home: Smart Student Living

Lydia Haack
Technical University of Munich

The studio for architecture and product design led by Prof. Richard Horden was approached in 2001 to look at the increasing problem that students have in finding affordable housing in Munich. A teaching programme began in collaboration with the Tokyo Institute of Technology, where students investigated the possibility of providing high-quality small-scale housing in a 2.55-metre cube. The intention was to provide housing for students that would form an enclosure for the four basic living functions – eating, sleeping, working and hygiene.

A further aspect was to study a modern lifestyle that is influencing society. Like a micro scooter, which is in between walking and cycling, the *i-home* explores the gap between product and house. It can easily be transported and relocated, and, like a car, can be 'parked' in a minimal space.

Very quickly, it became evident that the *i-home* project had great economic potential. Sites that normally have to be purchased might now be used on a temporary basis. The Munich student housing authority has a special interest in providing temporary housing projects, because they can now borrow public sites until they are finally developed. Hence, affordable student accommodation is now possible without cutting standards.

An *i-home* student village can be made available on demand very quickly and be adapted as student numbers change. In times of major events like athletics championships or the soccer World Cup 2006, these houses could be relocated and provide housing for athletes and/or press. A 'home', one of the most expensive consumer goods, can now be produced to allow changing 'fit-outs' to suit individual needs.

A proof of concept prototype was built in the Munich studio to check dimensional viability and challenge building regulations. The design focused on functional aspects such as precise zoning, flexibility, ease of use and changeability as well as spatial aspects. As in product design, attention was paid not only to the geometry but also to the other qualities such as touch. A set of *i-home* village components was developed to connect the inside and outdoor spaces in order to provide not only private but also public space.

The concept of the *i-home* is a volumetric harness allowing widely different cladding, window shapes and support systems. The fully developed product line will provide a range of models from 'high quality' with a car-like or airline-like standard of finish to 'basic', which will provide low-cost housing. The *i-home* can therefore be used for an extremely wide range of temporary housing needs. So far, it has attracted visitors and potential investors from as far away as Great Britain, Holland, South Africa and Australia. Currently under development is a prototype village for the Studentenwerk Munich, which will incorporate eleven *i-homes* for practical testing.

This project demonstrates a fresh approach to teaching, with both a professional and realistic approach to the planning and building process but also with an open mind and awareness of change. The *i-home* is the realization of a lightweight, minimized, transportable form of architecture that allows flexible use for both user and developer.

Project credits

Project collaboration with: Studentenwerk München, dipl- kfm. D. Maßberg
Teaching: R. Horden, L. Haack, H. Müller, W. Klasz
Students: Veronika Gruber, Claus Hainzlmeier, Stephan Koch, Taisi Tuhkanen, Bianca Matern, Daniel Oswald, Miroslav Penev, Vanessa Blacker-Sturm, Wiebke Seidler
Sponsors: Bund der Freunde TU Munich, Alcan Rolled Products, Alu-Meier, Munich, Gleixner Böhm, Design & Mechanik, Schaum-Stoffe Mayer Keller-Kalmbach Kvaddrat, Obi Heimwerkermarkt, Tek Enggruber

Portable Performance Space

George King
University of Liverpool

The brief for this project was to design a mobile performance space that would initially be used in and around Liverpool city centre, though it would subsequently travel to other venues around the world. The design could seat at least 150 people and contain a stage, backstage and foyer to allow theatre-type performances, but would also incorporate accommodation for corporate presentations, public speeches, etc. Most importantly the design would need to be fully operational within a few hours of arriving on site with a minimal number of operators.

The decision was made early on to use the trucks that would deliver the performance space to the site as part of the structure. This minimises the assembly time, as nothing needs to be 'unpacked'. The disadvantage is that that you are restricted in terms of the amount of floor space that you can fit onto the back of a 12.2-metre (40-foot) trailer.

The Portable Performance Space is in many ways primarily a technological solution to this problem. It was not just extra floor area that was needed but also extra height to create a dramatic space for the auditorium. The first response to the solution was a simple decision to use two trucks, one as the stage, the other as the seating. The second response was more innovative: to use layers of floors, roofs and walls that could 'fan' out to increase the dimensions of the building. The two halves of the building fit together so that although they are separate structures, they create a single space. A backstage area swings out from behind the stage and a foyer appears from below the seating.

This practical deployment concept utilises pneumatic rams to transform the trucks quickly and efficiently, creating a powerful and dynamic image to the public as the building is deployed. The building becomes part of the performance, a warm-up act before the main event takes place within.

This deployment concept was explored further in a peripatetic competition launched by the RIBA and RENEW (Regeneration Excellence in England's North West). The brief was to design a portable exhibition space that would be used to promote 'regeneration excellence in England's northwest'. Again the design was based upon a 12.2-metre (40-foot) trailer but this time it was refined to fit upon a single unit, though it would still triple the usable floor space once deployed.

The Portable Exhibition Space contains a perimeter gallery that surrounds a central office to allow constant visual contact for security purposes. The gallery space unfolds from one side of the trailer, using a similar deployment mechanism to that mentioned above, whilst ramp access and a large canopy unfold from the other side. The ramp not only conforms to Disability Discrimination Act regulations but also allows the building to 'touch' the landscape, not just to be an object in it. The canopy provides shelter whilst drawing the public in and advertising the event.

These two examples show how this basic deployment idea can be adapted to suit different buildings. The effect is not only dramatic and dynamic but has a practical use when the transportable building is required to be ready in minimal time and with a limited amount of human involvement. Because of this the possible uses are wide-ranging, from a concert venue to a field hospital.

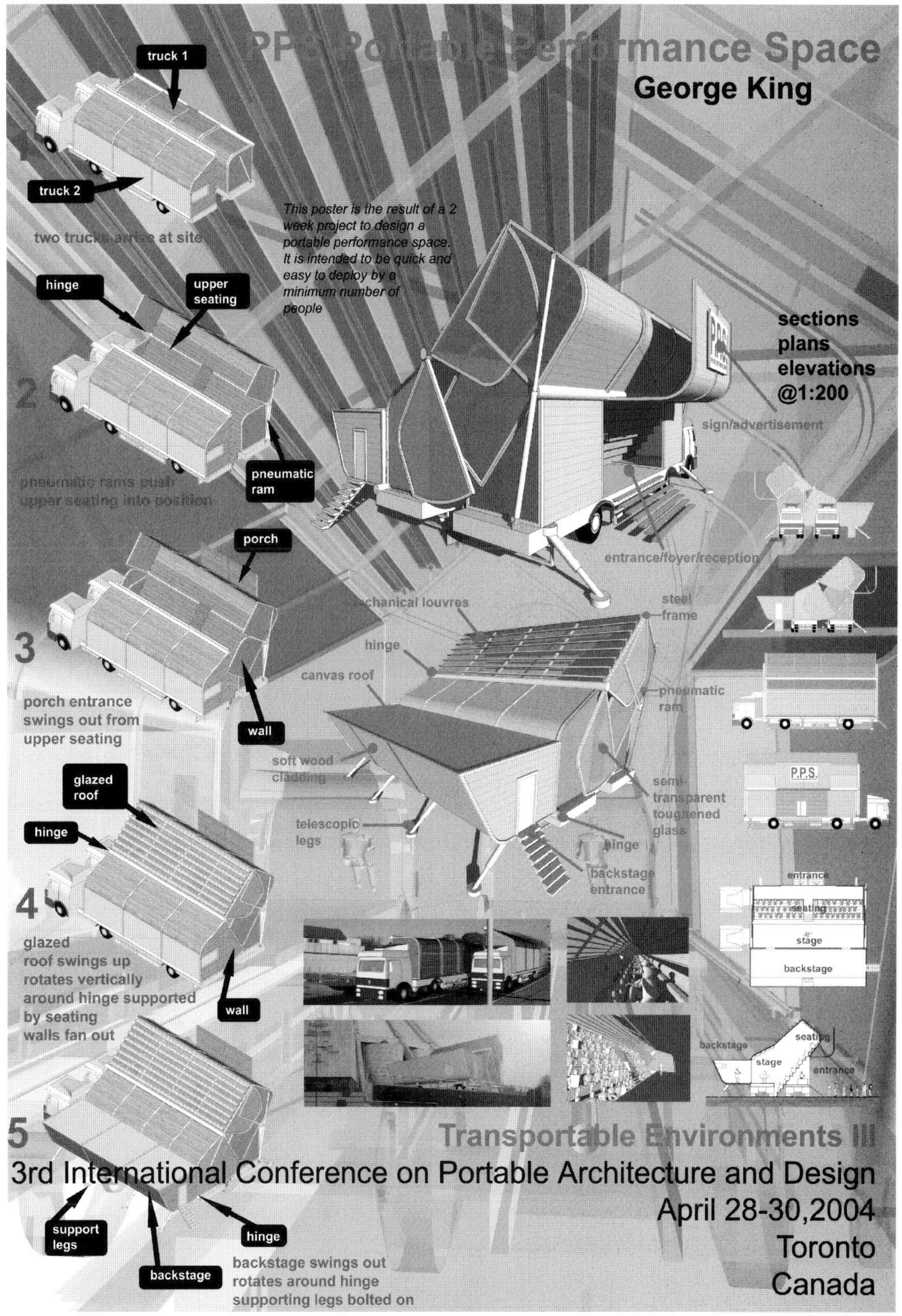

Mobile Clinic: A Transportable Treatment Unit for Sub-Saharan Africa

Piet Mazereeuw

One of the most important barriers in the world fight against HIV/AIDS is access to treatment (Médecins Sans Frontières, 2004). Modern science has made significant medical advances in the past decade but new drug treatment remains far beyond reach in certain parts of South East Asia, South America, Eastern Europe, the Middle East and Africa. The region of sub-Saharan Africa, with its vast and remote geography, is by far the most seriously affected, with approximately thirty million cases of HIV/AIDS.

In a recent article, entitled "How to Stop East Africa's Truckers From Driving the Aids Epidemic" from the Africa News Service, the International Transport Federation (ITF) states that:

> Mobility has been accepted by UN Aids and the International Labour Organization (ILO) as one of the biggest risk factors to infection with HIV. The findings of a demographic survey of truck drivers carried out in part of Kenya revealed that 84 per cent of the interviewees were suffering from some kind of sexually transmitted disease. In much of East Africa . . . the increase of commercial activity along transport corridors fuel a sex trade that services truckers, who are perceived as ready sources of scarce income. This, coupled with the fact that many of the transporters themselves have little knowledge in HIV prevention, often leads to unprotected encounters that result in the transmission of HIV.
> (Tiahmo, 2003)

Ahmed Omar, the secretary-general of the Kenya Long Distance Truck Drivers Association, says that his federation, in partnership with an American non-governmental organization, Africans United Against Aids Globally (AUAAG), launched a program to educate the long-distance drivers and their assistants against the risks of HIV on the highway. Already, the partnership has opened four information centers, housed in 6.1-metre and 12.2-metre (20- and 40-foot) containers, along the Mombasa–Nairobi transport corridor, where HIV awareness material is distributed free to crews of road haulage vehicles and other transporters.

The issues discussed above led to the design of a hybrid vehicle for the diagnostic treatment and prevention of HIV/AIDS. It was developed in coordination with Dr Maryan Mokobia and Dr Peter Ujomu from Health Matters Inc. in Lagos, and Joe Leberer from Médecins Sans Frontières in Toronto. This design proposal, the primary objective of which is to provide access to treatment for people in both urban and peri-urban areas, is a first step towards the establishment of a geo-medical healthcare network that is based on transportation infrastructure throughout the sub-Saharan region.

The design of the prototype is based on retrofitting the base model of the Mercedes Benz Vario 612 transport vehicle with medical space and equipment to function as a self-contained and fully operational mobile clinic. The interior and the exterior of the unit are programmed for three primary functions: preliminary diagnostics, anti-retroviral drug therapy and educational programming for different communities.

28.1 (Above) Mobile clinic and transportation node in Lagos (background image source: Edgar Cleijne, 2000); (Below left) Interior view of private examination room and light opening; (Below right) Educational slide of a microscopic view that illustrates HIV particles as projected on vehicle's side panel (Copyright Boehringer Ingelheim, 2004)

The unit also provides separate accommodation for four field doctors on the upper deck and is equipped with a screen for video projection on the vehicle's side panel and a medical information library in the front cab.

The body of the vehicle is mounted on a conventional rigid steel frame while the cab is molded as a fiberglass uni-body. In its parked position, the retractable components operate like a tent structure: poles are assembled, fabric is stretched and ramps are pulled out. The unit is completely independent with fuel-cell technology that generates sufficient power for lighting, heating, refrigeration, communications, and other day-to-day operations. Conventional automotive parts are purposely specified throughout the vehicle for easy replacement and interchangeability.

The built-in simplicity of the design takes into account the varying healthcare needs of the population in diverse regions of sub-Saharan Africa. With its on-road and off-road maneuverability, the unit can access dense urban areas such as food markets and motor parks, as well as isolated areas such as small villages and remote work camps. For fast and efficient mobilization, the treatment unit can be set up or taken apart in less than one hour by two qualified personnel. In case of medical emergency, the unit can be converted into a regional ambulance for the transportation of up to five patients to a nearby hospital and also be adapted to treat other diseases prevalent in the sub-Saharan region such as tuberculosis and malaria. Given the current disparity of healthcare systems in Sub-Saharan Africa, it is anticipated that this project may sponsor new economic possibilities for growth in the sector of mobile health services and public education.

> If we are able to halt the transmission of HIV along the region's highways, then we will be on our way to beating the killer virus across sub-Saharan Africa. What we need now is a fleet of cross-country vehicles to reach remote villages beyond the reach of our truck stops and motor parks.
> (Tiahmo, 2003)

References

Abdullahi, A. S. (2000) *Mobile Health Services and Health Service Delivery among Nomadic Populations in Northern Kenya*. Boston: Department of International Health.

Médecins Sans Frontières (2004) "Access to medicines at risk across the globe: What to watch out for in free trade agreements with the United States." Briefing Note, May. Retrieved from www.accessmed-msf.org/documents/ftabriefingenglish.pdf

The Mobile Health Clinic Project (n.d) "Health Help International in Zambia." Retrieved from home.freeuk.net/hhi/mobile_ambulance.htm.

Tiahmo, R. (2003) "How to stop East Africa's truckers from driving the Aids epidemic." Africa News Service, September 24.

Van Lerberghe, W., de Bethune, X., and de Brouwere, V. (1997) "Hospitals in sub-Saharan Africa: Why we need more of what does not work as it should," *Tropical Medicine International Health,* August 2(8): pp. 799–808.

Trajectory of the Junks

Janet McGaw
University of Melbourne

In systematic studies of the sustainability of urban environments, waste is one of the key flows to be analyzed (Yencken, 2000). However, as a program for urban design, the infrastructure and rituals surrounding garbage collection and disposal are rarely considered beyond sidewalk trash-can design. Three precedents of note inform this work, none specifically by architects. There are the transitory gardens and shelters built on Manhattan from reclaimed garbage by the city's homeless and photographically recorded by landscape architects Balmori and Morton (1993); architectural installation artist Tadashi Kawamata's vast and temporary splint built from reclaimed building waste around the quarantine hospital on New York City's Roosevelt Island, not exactly portable, but not permanent either (Gould, 1993); performance artist Mierle Laderman Ukeles' work as an unpaid artist in residence at the New York City Department of Sanitation, including the *Social Mirror* from 1983, a garbage collection truck clad in mirror, and a performance piece titled *Touch Sanitation* which involved a ritual of handshaking with 8,500 garbage collectors over a period of eleven months in the late 1970s (Gablik, 1991). All of these projects engage creatively with the spatial exercise of transportation and use garbage as both the object and primary material of their work.

Transportation and disposal are generally thought to be issues for council bureaucrats, garbage disposal businesses and civil engineers, not architects. How might architecture address not just the end point of garbage disposal but the processes of its transportation? Architecture is typically concerned with location on a fixed site rather than movement. And what might be the material nature of such an architecture? Traditionally materials have been thought to occupy two defined categories: the durable and the transient. According to sociologist Michael Thompson (1979), the third covert category is "rubbish" and has the status of maximum potential. Rubbish is defined not by its material durability or ephemerality, but by its flexibility. Discarded as valueless one day, it may be reborn the next.

New York City presents one of the most significant waste disposal problems of the western world. Fresh Kills Landfill on Staten Island, the repository of all its waste, is now the largest geographic feature on the 1,500-mile Atlantic seaboard from Florida to Maine. A decade ago, it was expected to be shut down by 2005, when it reached a height of 154 meters (505 feet) above sea level and impede flight paths into the city (Rathje and Murphy, 1992). Yet solutions to the problem have been beset by conflict between business and government.

Proposed here is a design solution that engages in the rituals surrounding collection, transportation, and ultimately incineration for the production of electricity in a plant located in a sunken island in New York City's upper bay, accessible only at high tide. It is a ceremonial process that is redemptive, converting waste into energy, yet also critical and poetic. Twice daily, garbage is to be delivered by municipal garbage trucks to "junks" at the end of the piers around Manhattan. These are floating membranes, towed by boats, which retract to form a tight ball when moving. The junks unfurl into the

(Left) Site plan Lowland of Redemption. Base mapped and edited from published map by US Geological Survey in 1967. Revised in cooperation with New York Department of Transportation in 1981. Title: Jersey City, N.J. – N.Y. N4037.5 – W7400/7.5. DMA 6165 II NE Series V822; (Top right) Perspective image showing built form of one of the electricity plants in the Lowland of Redemption; (Bottom right) Gondolier tending his junks; (Photomontage from unknown source)

sunken island, the "lowland of redemption," in a shallow area of the upper bay adjacent to Red Hook Channel when high tide renders the lip of the island accessible. This tripartite island, though permanent, is a vessel of moving parts. One third of the island ramps down, allowing the garbage to wash with the tide; the central portion has a curving and linear building housing an electricity plant that uses unsorted waste as a raw material. The Black Clawson process is an industrial design for a power plant that produces electricity by incinerating unsorted household waste at very high temperatures. It is currently in operation in various places around the world. The filtration pond for cleansing steam by-products will cyclically bubble into a formless, unpredictable architecture of sorts, like Bernard Tschumi's Fireworks, that "produces a delight which cannot be sold or bought" (Tschumi, 1975).

This design proposal demonstrates ways in which architecture can be made contingent on the movement and timing of natural and human processes. It shows how waste can be elevated to a new material status. A void in the bay and a fleet of transient boats can be juxtaposed with the skyscrapers of Manhattan as both a critique of the city and the site of its redemption.

"Trajectory of the Junks" *was a competition entry for USA Institute International Ideas Competition, Cyborg City: Artificial / Mechanical Islands for New York City's Rivers and Bay in 1999. It was placed second ex-equo.*

References

Balmori, D. and Morton, M. (1993) *Transitory Gardens, Uprooted Lives.* New Haven, Conn.: Yale University Press.

Gablik, S. (1991) *The Reenchantment of Art*. New York: Thames and Hudson, pp. 69–75.

Gould, C. (ed.) (1993) *Kawamata Project on Roosevelt Island*. New York: On the Table Inc., and Tokyo: Gendaikikakushitsu Publishers.

Rathje, W. and Murphy, C. (1992) *Rubbish! The Archaeology of Garbage*. New York: Harper and Collins, pp. 3–5.

Thompson, M. (1979) *Rubbish Theory: The Creation and Destruction of Value*. Oxford: Oxford University Press, p. 9.

Tschumi, B. (1975) "Questions of Space: The Architecture Paradox of the Pyramid and the Labyrinth." *Studio International*, September–October. Reprinted (1990) as "The Architectural Paradox: The Pyramid and the Labyrinth" in *Questions of Space: Lectures on Architecture*. London: AA Publications, p. 26.

Yencken, D. (2000) *Resetting the Compass: Australia's Journey Towards Sustainability,* Collingwood, Vic.: CSIRO Publishing.

Past Tents: A Scenographic Experiment

Kaija Vogel
Central St Martin's College of Art and Design

The visual vocabulary of *Past Tents* attempts to describe the cognitive sphere that represents and defines 'home'. It is part of an ongoing visual and design investigation intended to express *being* and our *basic human condition* in the spirit of Heidegger in order to define the 'phenomenology of home'. The challenge is to define home as something other than a stationary and physically enclosed space and consider it as a transportable environment that surrounds and moves with us. Involving three dancers, design and choreography of three tents, and light, sound and video installations, *Past Tents* uses Bertolt Brecht's *The Rise and Fall of the City of Mahagonny* as a narrative frame to define home as the interconnectedness of all things – a portable circular space.

Research began by designing a lightweight, sustainable and flexible tent, the form of which could be manipulated by folding, crunching, flipping, illuminating and stretching. The space of the tent was to represent absence of form while remaining simple, flexible and transportable. Raymond Abraham states that if: 'Architecture is essentially the conquest of place' then the tent becomes a 'formal manifestation of the physical absence [and/] or presence of man' (Brayer and Simonot, 2002). At times the tent demonstrated a conscious command of one's perimeter while intermittently indicating the loss of perception of one's environment. The presence of an inner or outer perimeter edge at the same time led to the creation of multiple spaces, allowing for further juxtapositions.

Two more tents were created and their relations with one another recorded. In one instance, the tents were conceived to be in the centre of a large urban green space in London where they were tossed and flipped by the wind for several hours, finally to be inhabited by children. The tents easily become a community and just as quickly they cease to be one. The study which began with only one basic single unit – one tent per person – expanded into the notion of a city and subsequently the rise and fall of a city.

Bertolt Brecht's *The Rise and Fall of the City of Mahagonny* was used to provide a narrative frame for three dancers, Keir Patrick, Ronnie Shapiro and Sapphire Williams, from the Laban Centre, London, UK, who were invited to collaborate on developing a vocabulary to communicate the idea of transportable space. The work developed into a series of cycles: the grand narrative (the city) and the sub-plots (the tents – their constant state of fugacity) and the dancers (the dwellers). Chris Baker was commissioned to compose a five-minute cyclical soundscape that built up from one single tone into multiple tones, slowly deconstructing back into one single note to continue the cycle. The tangible and intangible notions of home were represented by perimeters, boundaries and borders including various cycles and circles of music and motion.

Although the piece exists within a video timeframe as a whole (following several overlapping cycles), it can also function in a fragmented way as a series of visual and sonic haikus. Each moment can stand alone and be presented as a loop, as an on-going performance installation. In conclusion, the creative

process taken to devise *Past Tents* emphasizes the importance of flexibility, mobility and transformation in design and demonstrates the notion of 'home' as a cyclical space – a transportable environment which surrounds, drags behind or floats high above us.

Reference

Brayer, M. and Simonot, B. (eds) (2002) *ArchiLab's Future House: Radical Experiments in Living Space.* New York: Thames and Hudson.

Biographies

Martha Abbott is an adjunct assistant professor at the University of Minnesota. She received a B.A. degree from Stanford University and an M.Arch. from the University of Minnesota, with graduate studies in history and theory at the Architectural Association in London. She coordinates undergraduate design at Minnesota, and teaches first and second year design and theory courses. Her research interests lie in design theory and design education within the discipline of architecture.

Maziar Asefi received his Master's degree in architecture from the Iran University of Science and Technology (IUST) in 2001. He practised in Tehran from 1998 to 2001. He was a lecturer at Art University of Tabriz, 2001–03. He is currently working for a Ph.D. at the University of Liverpool, where he is a part-time tutor in the School of Architecture.

Justin Beal is an artist and architect based in New York and Los Angeles. More information about the project *Everything is Going According to Plan* and other projects is available at www.justinbeal.com.

Sarah Bonnemaison studied architecture at Pratt Institute, followed by a Master's degree at the Massachusetts Institute of Technology. After working on the design of deployable umbrellas for the great mosque of Mecca with Frei Otto and Bodo Rash in Stuttgart, she pursued her interest in festivals by completing a Ph.D. in cultural geography at the University of British Columbia. She now teaches architectural theory and the design of lightweight structures at Dalhousie University. She is also a member of a design practice, Filum Ltd, which focuses on temporary structures.

Prasad Boradkar is an assistant professor in the School of Design at Arizona State University in Tempe. Having studied mechanical engineering and industrial design in India, he received his graduate degree from the Ohio State University in Columbus, and has held positions at the Delft University of Technology in the Netherlands as well as at ITT Technical Institute in California.

Jordan Brandt received his degree in architecture from the University of Kansas in 2001, after which he received an Undergraduate Research Award to develop a carbon fibre medical cast prototype. This research ranged from digital imaging to CATIA training at the Boeing Corporation where he learned advanced surfacing and parametric programming. This experience took him to A. Zahner Co. in Kansas City, Missouri, where he worked as a design engineer on Gehry Partners' MIT Stata Center from 2001 to 2003, developing the undulating stainless steel/aluminium cladding system. He subsequently worked with Akro Fireguard as a design and materials engineer, prototyping and testing composite aircraft components. He received an AIA Fellowship for Advanced Research for his thesis 'Generation of Tertiary Systems: The Interstice of Art and Science' and is currently a doctoral student at the Harvard Graduate School of Design, pursuing research in construction metrology and freeform cladding systems.

Neil Burford is an architect and senior lecturer in the School of Architecture at the University of Dundee, where he was a founding member of an interdisciplinary research group of architects and engineers working in the field of lightweight structures. His principal areas of research are in lightweight and deployable structures, composite materials and the development of new environmentally benign biocomposite materials for use in construction.

John Chilton is a civil engineer and professor and head of school at the Lincoln School of Architecture. Previously he was a reader at the School of the Built Environment, University of Nottingham, where he taught for fifteen years. With a special interest in the history, design and construction of innovative and non-conventional structures, he is an active member of the International Association of Shell and Spatial structures (IASS). He sits on the IASS Executive Council and chairs the IASS Working Group 12, Spatial Wood Structures. Since 2000 he has been Scientific Coordinator for TensiNet, the European Union-funded network on tensile membrane structures.

Matias Creimer is a practising architect based in Los Angeles and teaches at OTIS College of Design. He moved to Los Angeles in 1994 to study with Daniel Libeskind and Thom Mayne. He received his Master's degree from UCLA in 1996. He is currently involved in the design and construction of numerous residential and commercial projects. His 'Dockable Dwelling' was among the winners of SECCA's HOME HOUSE international competition, sponsored by *Dwell Magazine* in 2003; it is published in Sean Topham's book *Move House* (Prestel, 2004).

Xenia F. Diente studied fine arts at the Cooper Union, New York. Her exploration of pattern, phenomena and metaphor began whilst she was studying architectonics with Manuel Baez. From 1998 to 2002, she worked for Chuck Hoberman, designing, illustrating and prototyping kinetic toys and structures as part of a team of mechanical engineers and artists. Currently, she is a public art manager and graphic designer at the NYC Department of Design and Construction. Xenia is attracted to the interrelation of art, architecture and engineering, finding moments where they meet on aesthetic and conceptual levels.

Andrew Furman is trained in architecture and interior design. He is a licensed interior designer and an assistant professor at Ryerson University's School of Interior Design where he coordinates the Spiral Lecture Series. As a member of the Pedestrian Space Forum he promotes pedestrian strategies in urban centres through public events that unite research and practice. His research explores movement systems in public space and their relationship to private space. Currently he is using photography as a means to explore sculptural views in the city.

Christoph Gengnagel is an architect and engineer and a senior assistant at the Lehrstuhl für Tragswerksplanung, Fakultaet für Architektur, Technical University of Munich. He has ten years' practical experience as a structural engineer and was involved in the realization of several membrane structures. His principal areas of research are in the design and analysis of membranes and deployable structures. A specific topic is the interactions between membranes and stiff structures.

Dean Goodman is a partner in Levitt Goodman Architects Ltd. This Toronto-based practice has won a number of awards during the last fifteen years including a Governor General Award for Excellence for 'Strachan House', a shelter for homeless adults constructed in an abandoned nineteenth-century factory in 1999. The firm's work includes projects in the not-for-profit sectors in housing and public institutions such as community centres and childcare facilities. Levitt Goodman is currently designing the new School of Architecture for the University of Waterloo in Cambridge, Ontario.

Renata Hejduk, Ph.D., is an assistant professor of architectural history and theory in the School of

Architecture and Landscape Architecture, College of Architecture and Environmental Design, Arizona State University. Her research is focused on avant-garde architecture and critical theory from around 1960 to the present. Her forthcoming book *Saved: The Religious Imagination in Modern and Contemporary Architecture*, co-edited with Jim Williamson, is to be published by Monacelli Press, NY.

Chuck Hoberman is an inventor of 'unfolding structures' – objects that transform their size and shape. His unique folding methods have been developed for uses ranging from miniature surgical tools to retractable stadium roofs. His recent and ongoing projects include design and production of toys, tents, emergency shelters, folding houses and portable theatres. For all of these varied uses, he makes structures that transform fluidly and completely. His projects have been shown at the 2002 Winter Olympics in Salt Lake City, Expo 2000 in Hanover, Germany, the California Museum of Science and Industry, Technorama der Schweiz, Switzerland, Museum of Modern Art, New York, and the Centre Georges Pompidou, Paris. He is currently the president of Hoberman Designs, Inc., which manufactures a full range of unique transforming toys based on his inventions. His work has featured in many publications including the *New York Times*, the *Wall Street Journal*, the *New Yorker*, *Wired*, *Architects' Journal*, *World Architecture*, *Architectural Record.* It has also been shown on television programmes such as *Today*, *Nightline*, CNN *News*, *Invention* on The Discovery Channel, *Tomorrow's World* on the BBC and *Dateline NBC*.

Filiz Klassen is an associate professor at Ryerson University. Her research focuses on the exploration of alternative, lightweight materials and fabrication methods for common design problems in built or made environments. She holds a Master of Architecture degree in Architectural Design from McGill University (1990) and a Bachelor of Architecture with honours from the Middle East Technical University (1987). She has given presentations at international conferences in Toronto, Singapore, London and New York and been a guest lecturer at Delft and Ankara. Her essays are published most recently in the book *Transportable Environments 2*, and in academic and professional journals such as the *International Journal of Art and Design*, *On Site*, and *Building* magazine. She is also an independent curator – her most recent work in this field is *Home: Mobile, Modular and Light*, which was displayed as part of *Digifest* at the Design Exchange Museum, Toronto, in May 2004.

Helmut Klassen is currently enrolled in the Ph.D. Joint Program in Communication and Culture at York and Ryerson Universities, Toronto, Canada. He has a post-professional Master's degree in the history and theory of architecture from McGill University, Montreal. He has a diverse academic and professional background that includes teaching history, theory and architectural design at both undergraduate and post-professional graduate levels at York University, McGill University, Carleton University, Ottawa, Canada, and the Middle East Technical University (METU) in Ankara, Turkey.

Elise Knudson graduated from the University of Wisconsin with a B.A. in philosophy and received a scholarship to study dance at the Alwin Nikolais and Murray Louis Dance Lab, New York. Her solo choreography has been presented at several New York showcases and festivals. Elise has had the pleasure of dancing for Wendy Perron, Chemeki/Lerner and J. Mandle Performance Group and she currently dances for Risa Jaroslow.

Robert Kronenburg, Ph.D. RIBA, is an architect and holds the chair of architecture at the University of Liverpool, School of Architecture, United Kingdom, where he is also the head of school. His books include *Houses in Motion: The Genesis, History and Development of the Portable Building*, *Portable*

Architecture, *Spirit of the Machine* and *Flexible Architecture*. Professor Kronenburg is a past Fulbright fellow and in 1998 received a visiting fellowship at St John's College, Oxford University. His research has received support from the Building Centre Trust, London, the Graham Foundation, Chicago, the UK Arts and Humanities Research Board, the Daiwa Foundation and the Leverhulme Trust. He curated the major exhibition *Portable Architecture* held at the Royal Institute of British Architects, London, and the touring exhibition *Spontaneous Construction* and he was curatorial adviser on the Vitra Design Museum's exhibition *Living in Motion*.

Vladimir Krstic is a professor of architecture at Kansas State University, USA. He was educated at the University of Sarajevo, and Kyoto University. While studying architecture in Japan he also worked and collaborated with Tadao Ando Architect and Associates. His scholarship is mainly focused on theoretical issues pertaining to the manifestation of the idea of the city in the conceptual structuring of architecture and its materialization. He has published extensively on this subject in relation to Japanese contemporary architecture in European and American journals and books.

LOT-EK is an ongoing investigation into 'artificial nature', or the unmappable outgrowth of familiar, unexplored, man-made and technological elements woven into the urban/suburban reality. LOT-EK extracts from this artificial nature prefabricated objects, systems and technologies to be used as raw materials. LOT-EK is the random encounter with such objects which are displaced, transformed and manipulated to fulfil programme needs. LOT-EK is the dialogue that develops with the specific features of these already existing objects thus generating unexpected spatial/functional solutions. LOT-EK is rethinking the ways in which the human body interacts with products and by-products of the industrial/technological culture. LOT-EK is reinventing domestic/work/play spaces and functions and questioning conventional configurations. LOT-EK is blurring the boundaries between art, architecture, entertainment and information. LOT-EK is an architecture studio based in New York. LOT-EK is Ada Tolla and Giuseppe Lignano, who were born and raised in Naples, Italy, graduated from the School of Architecture of the Università di Napoli (1982–89) and completed post-graduate studies at Columbia University in New York (1990–91).

Janet McGaw completed her M.Arch. by Design in 1999 and is currently working towards a Ph.D. by Creative Works, both at the University of Melbourne. Recent publications include 'Architectural (S)crypts: In Search of a Minor Architecture' in *Architectural Theory Review*, and 'Streetwise' presented at the Australasian Architecture Schools Association meeting in 2003. Recent creative works include *Bigger than a Black Hole and only 20cm Wide* in the RAIA exhibition *Small* in 2000 and *The Flâneur*, an urban installation included in the Faites de la Lumière in 2003. She maintains a small private practice and is a part-time teacher.

Alejandro Ogata attended the University of Kansas with a Fulbright/IIE grant and graduated with a Bachelor's in architecture. In 2001 he received the SOM travelling fellowship and lectured on his findings in the fields of architectural history, theory and technology upon his return. Ogata has been a visiting designer at Benetton's Fabrica research centre in Treviso, Italy, where he was offered a fellowship by the design department. Currently, Ogata is a LEED-accredited project designer for HNTB architecture in Kansas City, Missouri.

Marcin W. Padlewski established the Bakery Group Studio together with Michel DuVernet and Anissa Szeto in 1999. He holds a B.Arch. from the Carleton School of Architecture, Ottawa.

Alberto Pérez-Gómez, Ph.D., is Saidye Rosner Bronfman Professor of History of Architecture at

McGill University, Montreal, Canada, where he is director of Post-Professional (Master's and Doctoral) Programmes, and chairs the History and Theory of Architecture division. From 1990 to 1993, he was the founding director of the Institut de Recherche en Histoire de l'Architecture co-sponsored by the Canadian Centre for Architecture, the Université de Montréal and McGill University. He has taught at universities in Mexico City, Houston, Syracuse and Toronto, and at the Architectural Association in London, and has lectured extensively in North America and Europe. His numerous articles have been published in the *Journal of Architectural Education*, *AA Files*, *Arquitecturas Bis*, *Section A*, *VIA*, *Architectural Design*, *ARQ*, *SKALA*, *A+U*, *Perspecta* and many other periodicals. His first book in English, *Architecture and the Crisis of Modern Science* (MIT Press, 1983), won the Alice Davis Hitchcock Award in 1984. He is co-editor of the book series *Chora: Intervals in the Philosophy of Architecture* (McGill – Queen's University Press) and in 1997 co-authored with Louise Pelletier the book *Architectural Representation and the Perspective Hinge* (MIT Press). Dr Pérez-Gómez is currently engaged in a project to redefine the nature of architectural education by revisiting its historical sources during the Enlightenment and the early nineteenth century.

Patricia Pringle, AADip, M.Arch, trained as an architect in London where she practised before moving to Melbourne, Australia. She is a senior lecturer in the interior design programme at the Royal Melbourne Institute of Technology, School of Architecture and Design. She teaches design studies and coordinates the thesis year programme. Her current research explores the history of spatial amusements and their relationship with the new spatial disciplines of today.

Carolina Rodriguez qualified as an architect in Columbia and is currently studying for a Ph.D. based in the field of deployable structures at the University of Nottingham, United Kingdom. Her academic achievements include various international awards in the field of architecture and design, most recently 2nd Prize: Academic Laboratory to the Design of Public (Colombia 1998), 1st Prize: Competition 'Otto De Greiff' (Colombia. 2001), 3rd Prize: 7th Student Competition Textile Structures For New Building (Germany 2003) and Hangai Prize IASS (Taiwan 2003). She is the author of *Arquitectura Metamorfica* (2000).

Jennifer Siegal is the principal and founder of the Office of Mobile Design, a progressive architecture/design studio that is dedicated to the exploration and production of mobile and eco-logic structures. She earned a Master's degree from the Southern California Institute of Architecture (SCI-Arc) in 1994, and was a 2003 Loeb Fellow at Harvard University's School of Design where she explored the use of intelligent, kinetic and lightweight materials. Ms Siegal's work was exhibited at the Cooper Hewitt, National Design Museum's 2003 National Design Triennial: *Inside Design Now*; and the Walker Art Center's *Strangely Familiar: Design and Everyday Life*. In 2003, *Esquire* magazine named her one of the 'Best and Brightest' and the Architectural League of New York included her in the acclaimed Emerging Voices programme. Ms Siegal is a Full Professor at Woodbury University in Los Angeles, and the editor of *Mobile: The Art of Portable Architecture*. Her forthcoming monthly publication series entitled *Materials Monthly* will be launched in 2005.

Meindert Versteeg received his degree in interior architecture from the Royal Academy of Arts, Department of Architectural Design at The Hague in 1998. His work focuses on design for extreme living and working conditions, mobile and transformable environments, and work/live space modules. Over the last few years he has received several grants from the Dutch Foundation of Art, Design and Architecture, which allowed him to gain working experience in this field of architecture, with opportunities to research small living spaces in Tokyo and mobile homes in California. In 2003 he published two papers describing a concept of a mobile rover/habitat designed for Mars. Recently he has been invited to participate in the space habitation study 'SpaceHeaven' that explores habitats for outer space in the near future and in forthcoming planetary missions.

Kaija Vogel is an artist and designer based in London, UK. Trained as a fine artist and filmmaker in Canada and the UK, she holds an MA in scenography from Central St Martin's College of Art and Design, London. Inspired by ideas and concepts of temporary and mobile homes, her work includes multimedia installation, video, performance and sculpture illustrating social, semiotic, urban and architectural notions of 'home'. As well as being a multimedia artist, Kaija works as a production designer and art director in film and television. Her most recent installation work *Grounding Part I and Part II* (exhibited in London, December 2003) is a series of ten sculpture pieces incorporating air craft alloy, plaster and video.

Selected Bibliography

Adorno T. (1976) *Introduction to the Sociology of Music*. New York: Continuum Press.

Adorno, T. (1991) *The Culture Industry: Selected Essays on Mass Culture*. London: Routledge.

Alvarez, A. (1987) *Offshore: A North Sea Journey*. Paris: Flammarion.

Ambasz, E. (1972) *Italy: The New Domestic Landscape. Achievements and Problems of Italian Design*. New York: The Museum of Modern Art in collaboration with Centro Di, Florence.

Antonelli, P. (1995) *Mutant Materials in Contemporary Design*. New York: MOMA, distributed by Abrams.

Appadurai, A. (1986) 'Introduction: Commodities and the Politics of Value'. in A. Appadurai (ed.), *The Social Life of Things: Commodities in Social Perspective* (pp. 3–63). Cambridge: Cambridge University Press.

Arthur, C. (2003) 'Mr. iMac Wins Design Prize for Banishing Beige Boxes', *Independent*, 3 June, p. 7.

Ashby, M. and Johnson, K. (2002) *Materials and Design: The Art and Science of Material Selection in Product Design*. Oxford: Butterworth-Heinemann.

Balmori, D. and Morton, M. (1993) Transitory Gardens, Uprooted Lives. New Haven and London; Yale University Press.

Banham, Mary *et al*. (1996) *A Critic Writes: Essays by Reyner Banham*. Berkeley: University of California Press.

Béguin, F. (2001) 'Offshore Oil: The Sixth Continent'. in *Quaderns #18: Architecture and Energy.*

Béret, C. (ed.) (1996) *Frederick Kiesler: Artiste-Architecte.* Paris: Editions du Centre Georges Pompidou.

Berger, P. (1997) *Redeeming Laughter: The Comic Dimension of Human Experience*. New York: Walter de Gruyter.

Beukers, A. and Hinte, E. V. (2001) *Lightness: The Inevitable Renaissance of Minimum Energy Structures*. Rotterdam: 010 Publishers.

Blaser, W. (1999) *Werner Sobek: Art of Engineering*. Basel: Birkhauser.

Bogner, D. (1988) *Frederick Kiesler: Architekt, Maler, Bildhauer, 1890–1965*. Vienna: Locker.

Bogner, D. (1997) *Frederick Kiesler: Inside the Endless House*. Vienna: Bohlau.

Bokin, J. and Lukes, T. (1994) *Marcuse from the New Left to the Next Left*. Lawrence: University of Kansas Press.

Braddock, S. and O'Mahony, M. (1999) *Techno Textiles: Revolutionary Fabrics for Fashion and Design*. New York: Thames and Hudson.

Brayer, M. and Simonot, B. (eds) (2002) *Archi-Lab's Future House: Radical Experiments in Living Space.* New York: Thames and Hudson.

Breen, C. (2002) *Secrets of the iPod*. Berkeley: Peachpit Press.

Buchanan, P. (1997) *Renzo Piano Building Workshop: Complete Works.* London, Phaidon.

Bucquoye, M. (ed.) (2002) *Van Bakeliet tot Composiet: Design met Nieuwe Materialen, (From Bakelite to Composite: Design in New Materials).* Ghent: Design Museum, Stichting Kunstboek.

Bull, M. (2000) *Sounding Out The City: Personal Stereos and the Management of Everyday Life*. Oxford: Berg.

Bull, M. (2004) 'To Each Their Own Bubble': Mobile Spaces of Sound in the City', in A. Couldry and A. McCarthy (eds), *Mediaspace: Place, Scale and Culture in a Media Age* (pp. 275–293). London: Routledge.

Burford, N. and Gengnagel, C. (2003) 'Ein Dach der besonderen Art', in *Mitteilungen der TUM 03/04, Munich*.

Burkhart, B. and Hunt, D. (2000) *Airstream: The History of the Land Yacht*. San Francisco: Chronicle Books.

Candela, F., Perez, P. E., Calatrava, S., Escrig, F. and Perez, V. J. (1993) *Arquitectura Transformable*, Seville: Escuela Superior de Arquitectura de Sevilla, Spain.

Cartwright, L. (1995) *Screening the Bod: Tracing Medicine's Visual Culture.* Minneapolis: University of Minnesota Press.

Chilton, J. C., Choo, B. S. and Wilkinson, D. (1998) 'A Parametric Analysis of the Geometry of Retractable Reciprocal Frame Structures' in *Light-weight Structures In Architecture, Engineering and Construction*, Vol. 1, pp. 547–555.

Constant, C. (2000) *Eileen Gray.* London: Phaidon.

Cook, P. (1970) *Experimental Architecture*. New York: Universe Books.

Cook, P. *et al.* (1972) (1999) *Archigram*. Originally published: Basel: Boston: Birkhauser. Republished: New York: Princeton Architectural Press.

De Landa, M. (1997) *A Thousand Years of Linear History*. New York: Zone Books.

De Zegher, C. and Wigley, M. (eds) (2001) *The Activist Drawing: Retracing Situationist Architectures from Constant's New Babylon to Beyond.* New York: The Drawing Center.

Deleuze, G. (1977) *Dialogues/Gilles Deleuze, Claire Parnet*. Paris: Flammarion.

Deleuze, G. and Guattari, F. (1987) *A Thousand Plateaus: Capitalism and Schizophrenia* (B. Massumi, Trans.). Minneapolis: University of Minnesota Press.

DeNora, T. (2000) *Music in Everyday Life*. Cambridge: Cambridge University Press.

Dessauce, M. (1999) *The Inflatable Moment: Pneumatics and Protest in '68*. New York: Princeton Architectural Press.

Devant, D. (1909) *Magic Made Easy.* London: Cassell.

du Gay, P., Hall, S., Janes, L., Mackay, H. and Negus, K. (1997) *Doing Cultural Studies: The Story of the Sony Walkman*. London: Sage Publications.

Duchamp, M. (1973) *The Writings of Marcel Duchamp*. New York: Da Capo Press.

Eaton, R. (2001) *Ideal Cities: Utopianism and the (Un)Built Environment*. London: Thames and Hudson.

Edmondson, A. (1987) *Fuller Explanation: The Synergetic Geometry of R. Buckminster Fuller. Boston:* Birkhäuser.

Faegre, T. (1979) *Tents: Architecture of the Nomads*. New York: Anchor Books.

Fawcett, C. (1980) *The New Japanese House: Ritual and Anti-Ritual, Patterns of Dwelling*. New York: Harper & Row.

Fitchen, J. (1989) *Building Construction Before Mechanization*. Cambridge, Mass.: The MIT Press.

Foucault, M. (1986) 'Of Other Spaces', *Diacritics 16* (Spring 1986), pp. 22–27.

Frampton, K. (2001) *Studies in Tectonic Culture*. Cambridge, Mass.: The MIT Press.

Franklin, R. M. (1978) *Beyond Metabolism: The New Japanese Architecture*. New York: Architectural Record Books.

Gablik, S. (1991) *The Reenchantment of Art*. New York: Thames and Hudson.

Gantes, C. (2000) *Deployable Structures Analysis and Design*, Billerica, Mass.: Wit Press.

Giedeon, S. (1948) *Mechanization Takes Command: A Contribution to Anonymous History*. New York: Norton.

Hall, E. T. (1990) *The Hidden Dimension.* Toronto: Anchor Books.

Hanaor, A. and Levy, R. (2001) 'Evaluation of Deployable Structures for Space Enclosures' *International Journal of Space Structures*, 16(4), pp. 211–230.

Hay, H. [1950] (1982) *The Amateur Magician's Handbook*, rev. edn New York: Harper and Row.

Held, R. L. (1982) *Endless Innovations: Frederick Kiesler's Theory and Scenic Design.* Ann Arbor, Mich.: UMI Research Press.

Henderson, L. D. (1998) *Duchamp in Context: Science and Technology in the Large Glass and Related Works.* Princeton, NJ: Princeton University Press.

Hernandez, C. and Zalewski, W. (1993) 'Expandable Structure for the Venezuelan Pavilion at Expo '92' in G. A. R. Parke and C. M. Howard (eds), *Space Structures* Vol. 2, pp. 1710–1719. London: Thomas Telford.

Herwig, O. (2003) *Featherweights: Light, Mobile and Floating Architecture*. Munich: Prestel.

Higgins, J. (2001) *The Raymond Williams Reader*. Oxford: Blackwell.

Highmore, B. (2002) *Everyday Life and Cultural Theory: An Introduction*. London: Routledge.

Hillier, B. and Hanson, J. (1993) *The Social Logic of Space*. Cambridge: Cambridge University Press.

Hoberman, C. (1996) 'Temporary Unfolding Structures' in *Detail; Temporary Structures*, Munich.

Horden, Richard (1995) *Light Tech: Towards a Light Architecture*. Boston: Berkhäuser.

Joseph, F. and Look, J. (2003) *Tadashi Kawamata, Bridge and Archive*. Bielefeld: Kerber Verlag.

Kachur, L. (2001) *Displaying the Marvelous*. Cambridge, Mass.: The MIT Press.

Kieran, S. and Timberlake, J. (2004) *Refabricating Architecture: How Manufacturing Methodologies are Poised to Transform Building Construction*. New York: McGraw Hill.

Kiesler, F. (1996) *Frederick J. Kiesler: Selected Writings.* Stuttgart: Verlag Gerd Hatje.

Kikutake, K. (1959) Marine Cities. *Kokusai Kenchiku* (January).

Kostof, S. (1999) *The City Assembled. The Elements of Urban Form Through History*. New York: Bullfinch Press.

Krell, David (ed.) (1993) *Martin Heidegger, Basic Writings*. London: Routledge.

Kronenburg, Robert (1997) *FTL: Softness, Movement and Light*. London: Academy Editions. Architectural Monographs No. 48.

Kronenburg, Robert (2001) *Spirit of the Machine: Technology as an Inspiration Architectural Design*. London: John Wiley.

Kronenburg, Robert (2002) *Houses in Motion: The Genesis, History and Development of the Portable Building* (2nd expanded edn, 1st edn 1995), London: John Wiley.

Kronenburg, Robert (2003) *Portable Architecture* (3d edn), Oxford: Architectural Press.

Kronenburg, Robert (ed.) (1998) *Transportable Environments: Theory, Context, Design and Technology*. London and New York: Routledge.

Kronenburg, Robert, Lim, Joseph and Wong, Y. C. (eds) (2002) *Transportable Environments 2*. London and New York: Routledge.

Kuenzli, R. F. N. E. (ed.) *Marcel Duchamp: Artist of the Century*. Cambridge, Mass.: The MIT Press.

Kunkel, P. (1999) *AppleDesign: The Work of the Apple Industrial Design Group*. New York: Graphis.

Lang, P. and Menking, W. (2003) *Superstudio: Life without Objects*. Milan: Skira Editore S.p.A.

Langbecker, T. (1999) 'Kinematic Analysis of Deployable, Scissor Structures', in *International Journal of Space Structures*, 14 (1).

Latour, B. (1986) 'Visualization and Cognition: Thinking with Eyes and Hands'. in *Knowledge and Society* 6: 1–40.

Leach, Neil (1999) *The Anaesthetics of Architecture*. London: MIT Press.

LeCuyer, A. (2003) *Steel and Beyond: New Strategies for Metals in Architecture*. Basel: Birkhäuser.

Lee, P. M. (2000) *Object to Be Destroyed: The Work of Gordon Matta-Clark.* Cambridge, Mass.: The MIT Press.

Lefebvre, H. (1999) *Everyday Life in the Modern World*. New Brunswick: Transaction.

LOT-EK (2002) *LOT-EK: Urban Scan*, London: Lawrence King Publishing.

Manzini, E. (1989) *Material of Invention*. Cambridge, Mass.: MIT Press.

Marcuse, H. (1964) One-dimensional Man: Studies in the Ideology of Advanced Industrial Society. Boston: Beacon Press.

Marcuse, H. (1969) *Essay on Liberation*. Boston: Beacon Press.

Marshall, Andrew C. (1988) *Composite Basics-5*. Walnut Creek: Marshall Consulting.

Marwick, A. (1998) *The Sixties: Cultural Revolution in Britain, France, Italy and the U.S. 1958–1974*. New York: Oxford University Press.

Mellor, James (ed.) (1970) Buckminster Fuller 'Designing a New Industry' in *The Buckminster Fuller Reader*. London: Jonathan Cape.

Messerschmidt, E. and Bertrand, R. (1999) *Space Station, Systems and Utilization*. Munich: Springer.

Mitchell, W. (2002) 'E- Bodies, E- Building, E- Cities'. in N. Leach (ed.), *Designing for a Digital World*. (pp. 50–56). Chichester: Wiley-Academy.

Mori, T. (ed.) (2002) *Immaterial/Ultramaterial: Architecture, Design, and Materials*. New York: Harvard Design School in Association with George Braziller Publisher.

Ockman, J. (1993) *Architecture Culture 1943–1968: A Documentary Anthology.* New York: Columbia Books of Architecture/Rizzoli.

Otto, F. and Rasch, B. (1995) Finding Form, Towards an Architecture of the Minimal, Catalogue for the Exhibition in the Villa Stuck. Munich.

Pearson, D. (2001) *Yurts, Tipis and Benders: The House that Jack Built.* London: Gaia Books.

Pérez-Gómez, A. (1992) *Polyphilo, or, the Dark Forest Revisited: An Erotic Epiphany of Architecture*. Cambridge, Mass.: MIT Press.

Pettena, G. (1982) *Superstudio: 1966–1982. Storie, Figure, Architettura*. Florence: Electa Firenze.

Phillips, L. (ed.) (1989) *Frederick Kiesler*. New York: Whitney Museum of American Art in association with W. W. Norton.

Ponticel, Patrick (2003) 'Boeing Takes a Leap Forward with Composites' in *SAE Aerospace*, November, pp. 29–31.

Popovic, O., Chilton, J. C. and Choo, B. S. (1997) 'The Variety of Reciprocal Frame (RF) Morphologies Developed for a Medium Span Assembly Building-Case Study' in J. C. Chilton *et al.* (eds) *Structural Morphology: Towards the New Millennium*, pp. 164–171. University of Nottingham.

Reux, Frédéric (2004) 'Composites in Buildings: Promising Dynamics', in *JEC Composites*, pp. 7, 28–30.

Rodriguez, C. (2000) *Arquitectura Metamórfica*, Bogota: ICFES and National University of Colombia.

Rodriguez, C. and Chilton, J. (2003), 'Swivel Diaphragm a New Alternative for Retractable Ring Structures', *Journal of the International Association for Shell and Spatial Structures*, Vol. 44 n. 3, pp. 181–188.

Rosak, T. (1968) *The Making of a Counter Culture: Reflections on the Technocratic Society and Its Youthful Opposition*. Berkeley: University of California Press.

Rowe, C. and Slutzky R., (1964/1997) *Transparency.* Basel: Birkhäuser Verlag.

Sadler, S. (1998) *The Situationist City.* Cambridge, Mass.: The MIT Press.

Schutz, A. (1973) *Collected Papers 1: The Problem of Social Reality.* The Hague: M. Nijhoff.

Scoates, Chris (ed.) (2003) *LOT-EK-MDU.* Santa Barbara: Walker Art Center and the Art Museum of the University of Santa Barbara, California.

Sennett, R. (1974) *The Fall of Public Man*. New York: Norton.

Shan, W. (1993) 'Configuration Studies of Foldable Structures' in G. A. R. Parke and C. M. Howard (eds), *Space Structures 4*, Vol. 1, pp. 824–832. London: Thomas Telford.

Sharpe, S. (1985) *Conjurors' Optical Secrets*. Calgary: M. Hades International.

Siegal, Jennifer (ed.) (2002) *Mobile: The Art of Portable Architecture*, New York: Princeton Architectural Press.

Simmel, G. (1997) 'The Metropolis and Mental Life', in D. Frisby and M. Featherstone (eds), *Simmel on Culture: Selected Writings*, pp. 174–186. London: Thousand Oaks.

Stoos, T. (ed.) (1994) *A Guide to Archigram: 1961–74*. London: Academy Editions.

Strong, A. Brent (1989) *Fundamentals of Composites Manufacturing: Materials, Methods and Applications.* Dearborn: Society of Manufacturing Engineers.

Topham, S. (2002) *Blowup: Inflatable Art, Architecture and Design*. Munich: Prestel.

Tsutomu, K. and Tokai, H. (1997), 'Cable Scissors Arch-Marionettic Structure' in J. C. Chilton *et al.* (eds) *Structural Morphology: Towards the New Millennium*, pp. 107–114. University of Nottingham.

Tzonis, A. (1999) *Santiago Calatrava: The Poetics of Movement*. London: Thames and Hudson.

Vincent, J. F. V. (2001) 'Deployable Structures in Nature', *Courses and Lectures – International Centre for Mechanical Sciences* (412), pp. 37–48.

Wan Ho, M. (1997) 'The New Age of the Organism', *Architecture Design Profile No. 129. New science- New Architecture,* Vol. 67, No. 9/10.

Wilcoxon, R. [1968] (1976) 'Council of Planning Librarians Exchange Bibliography', in R. Banham, *Megastructure: Urban Futures of the Recent Past*. London.

Wilk, C. (1981) *Marcel Breuer: Furniture and Interiors.* London: Architectural Press.

Williams, R. (1975) *Television: Technology and Cultural Form*. New York: Schocken Books.

Winnicott, D. W. (1971) *Playing and Reality.* London: Tavistock Publications.

Winnicott, D. W. (1986) *Home is Where We Start From: Essays by a Psychoanalyst.* New York and London: Norton.

Yahya, H. and Clarke, A. (2002)*The Design in Nature*. London: Ta- Ha Publishers.

Index

Titles of works of art or literature are given in italics. Names of oil rigs are given in italics